The Guinea Fowl (*Numida meleagris*)

José del Moral de la Vega
Jerónimo del Moral Martínez
Editors

The Guinea Fowl (*Numida meleagris*)

Integrated Management for Meat Production and Pest Control

Springer

Editors
José del Moral de la Vega
Emeritus Head of the Department
of Plant Pathology of CICYTEX, (Scientific
and Technological Research Centre
of Extremadura)
CICYTEX-Finca La Orden
Guadajira, Badajoz, Spain

Jerónimo del Moral Martínez
Food Quality and Microbiology Research
Group (AGA015) of the Extremadura
Science, Technology and
Innovation System,
(Badajoz), Spain

Translation from the Spanish language edition: "La Pintada (*Numida meleagris* L.): un ave para la biodiversidad y el sostenimiento del mundo rural" by José del Moral de la Vega and Jerónimo del Moral Martínez,

ISBN 978-3-032-11653-6 ISBN 978-3-032-11654-3 (eBook)
https://doi.org/10.1007/978-3-032-11654-3

This Springer imprint is published by the registered company Springer Nature Switzerland AG
The registered company address is: Gewerbestrasse 11, 6330 Cham, Switzerland

To agricultural engineers Víctor Moreno Márquez and Antonio Arias Giralda, whose research on locusts Dociostaurus maroccanus (Thunberg, 1815) has been invaluable in controlling plagues of this insect.

To biologists Ramón de Arcos Nieto–Guerrero, José González Crespo and Pedro Gómez Hernández for their contributions to the first studies on locust predation by the guinea fowl (Numida meleagris L.).

Preface

In 1988, a group of researchers from various European countries took part in a meeting on phytotechnology organised by the CIHEAM—Centre International de Hautes Études Agronomiques Méditerranéennes (International Centre for Advanced Mediterranean Agronomic Studies). The meeting was chaired by a politician from Brussels who explained to us in detail that the countries of the European Union would have to reduce agricultural activity. His reasoning was convincing: people who base their economies on the secondary sector (industry) or the tertiary sector (trade, tourism, finance, etc.) are more developed than those who base their economies on the primary sector (agriculture, fishing, mining, etc.), so our activity in the primary sector should be drastically reduced. On the other hand, if Europe wanted to sell industrial products and services to Third World countries, there would be no choice but to let them take over a large part of the activity that we developed in the primary sector and then buy the products generated from them, with the sale of which these countries could pay us what we sold them. The reasoning seemed correct, and the immediate consequence of this was clear: in the EU, the area dedicated to crops and livestock would have to decrease and, consequently, European farmers would have to reduce their activity.

In 2008, the goal set by the European Union seems to have been achieved. A study by the Humboldt University of Berlin on land use for agricultural production in the European Union states: since 1999, net imports have grown by 40% in the EU, which means that 35 million hectares of land have been virtually occupied—outside our borders—to produce the grains we import.

But in Europe, as in so many other parts of the world, there are regions where agricultural activity is fundamental; parts of the central, southern and southwestern Iberian Peninsula, among others, and as this activity has been reduced as a result of political measures—such as subsidies to reduce cultivation areas or subsidies at the margin of production—and not as a product of social evolution, a peculiar situation has arisen. There are counties where the wealth generated by agricultural activity, not being now generated by industry or private enterprise services, is being taken over by the various administrations (local, regional, state and/or supranational) through subsidies for activities which, in many cases, are of dubious utility. This, in turn, generates generalised disillusionment that discourages young people, the final consequence of which is emigration and the emptying of entire villages which are left inhabited, for the most part, by a population of elderly people. At present, the southern countryside of Extremadura (Spain), one of the regions that can be taken as an example of the above, has a population density of only 13 inhabitants per square kilometre and a negative vegetative growth, with over 20% of houses uninhabited from which it can be expected that, within a few years, if this situation is not remedied, the emptying of villages will begin in this region, a situation which, for various other reasons, is repeated all over the world.

Certain groups of opinion have thought that the solution to this problem would be to let the population continue to abandon these purely agricultural towns to regroup in the more industrialised cities; but if this were to happen, it would mean that many small towns, villages and farmhouses would remain empty. The same opinion groups have suggested, as a positive ecological measure, that abandoned agricultural land could be used for forestry areas, areas which, in many cases and spontaneously, would be occupied by native woodland. But this approach does not seem to be feasible because of the particularity of the rural areas where the problem exists.

One of the milestones of civilisation is the beginning of agriculture, and from the discoveries made in Abu Hureyra (Syria), we now know that this phenomenon took place in Eurasia 11000 years ago in an area around the Tigris and Euphrates rivers—overlapping and sharing plant species with the American continent, which places the origin of agriculture in the area of the Andes—just at the eastern end of the Mediterranean. The special characteristics of this sea—it has four peninsulas and 40% of its coastline is made up of islands—make it ideal for communication between peoples, which is why agriculture expanded rapidly in the casserole formed by the countries bordering it from Mesopotamia to the Iberian Peninsula. In these areas bordering the Mediterranean Sea—the sea in the middle of the land (meaning of Mediterranean)—one of the potteries of our current civilisation was located, and for this reason the European regions of these areas are like an immense museum with vestiges of the cultures that, for 11000 years, have succeeded each other (Sumerians, Assyrians, Egyptians, Phoenicians, Persians, Greeks, Romans, etc.).

It could be envisaged that custody of this heritage would be carried out in these ghost towns by specialised companies, or with public employees, from cities located in strategic places; but this does not seem feasible. Creating industries in these places, even if they were derived from agricultural products, would serve to create wealth, but it would hardly avoid the problem of the brutal depopulation that is occurring in these areas. And programming, as a political ambition, the creation in this rural world of a "Silicon Valley" with biotechnology industries and research companies, would be a solution far removed from reality.

For all of the above reasons, there seems to be no other way forward than to reintroduce agriculture to these rural populations of the world, albeit “in new clothes”.

We need to implement an agriculture with more productive elements of biological than industrial origin, aimed at generating more excellence than vulgarity, structured in harmony with tourism, based more on innovation than on routine, constituted more as a business than as a "farm". An agriculture that generates sufficient benefits so that the people who live from it have the same income and enjoy the same level of social welfare as those who live from the secondary and tertiary productive sectors. The day that a young person feels happy to work on a farm and thinks that his or her life has a future developing this activity will be an unequivocal sign that the growing depopulation of the rural world has been halted and, consequently, the exodus to the peri-urban belts of the big cities has been halted. Reaching that day does not seem to be easy or likely to happen in the short term, and it must be sustainable.

The development of agriculture throughout history has led to a reduction in the number of plants and animals used in agriculture. Do many farmers today know what are the azerion, kadi, huchaki, etc., cultivated by the Arabs in the Aljarafe in Seville (Spain) in the twelfth century? What remains of those dovecotes that existed in all rural houses until the middle of the twentieth century? And what about the rabbits, turkeys, chickens and other animals that were kept in almost every farmhouse? The improvement of agriculture has gradually wiped out many of them, a phenomenon that has been exacerbated dramatically since the so-called Green Revolution. In China, of the 10000 varieties of wheat that existed in 1949, only about 1000 remained in the 1970s. Of the 7000 varieties of apple trees counted in the United States in the nineteenth century, only 14% remain in the twenty-first century. The problem is to be solved by recovering plant and animal species, varieties and breeds, in order to preserve them from the genetic erosion caused by agricultural over-productivity.

Columela (first century) cited, among the volatile birds present in Spanish agriculture of his time, a screaming species with spots in its plumage that farmers called the guinea fowl; when in the mid-twentieth century, the countryside was still full of farmhouses and farmsteads, this bird accompanied the chickens, ducks and turkeys that roamed around the farmhouses. Could this species be recovered to integrate it into the production systems of sustainable agriculture? This was one of the objectives that a team of Spanish researchers undertook in 1995.

Locust plagues have been the paradigm of misfortunes that, like a curse, appeared in the countryside to cause misery in the rural world. Until the mid-twentieth century, in many areas of the Mediterranean, they were controlled by fire, manual destruction of the insect or ploughing. The discovery of chlorinated insecticides and their application against this insect, from the 1950s onwards, reduced the areas affected to certain areas where the insect is endemic, but at what cost: at the cost of applying thousands of tonnes of insecticides. First chlorinated (DDT, HCH), then phosphorous (fenitrothion and malathion), and now pyrethroids (deltamethrin) and biorational insecticides (diflubenzuron), which means that the control of these plagues indirectly pollutes the environment.

The control of locust infestations with insecticides is well justified, but this produces a sad paradox, and that is that this procedure is probably causing the loss of one of the greatest riches of these poor regions, a richness which is their primitivism, an extraordinary quality factor for the agricultural products produced there. For this reason, it seemed necessary to start looking for a new procedure for biological control of locusts, which would have little or no impact on the environment and would be as effective as synthetic insecticides. Predation was chosen as one of these methods. And if the predatory animal was of interest to human food, it would eat the locust and man would eat the predator, transforming, in this elementary way, a calamity into a source of wealth.

To determine this biological control procedure by means of predation, a preliminary study was carried out, starting with four species of birds: the turkey (*Meleagris gallopavo* L.), the partridge (*Alectoris rufa* L.), the hen (*Gallus gallus* L.) and the guinea fowl (*Numida meleagris* L.), choosing the guinea fowl because of the good results obtained in the experiments carried out. Two research projects were carried out with this predatory bird, financed by the Spanish INIA (National Institute for Agricultural Research), ERDF (European Regional Development Fund) and the Regional Government of Extremadura (Spain), proving that locust infestations can be controlled by using guinea fowl kept in elementary facilities and in almost total freedom. The experiments not only demonstrated the efficacy and economic feasibility of the process but also showed that guinea fowl feeding spontaneously and ad libitum on locusts, other arthropods and grains had excellent organoleptic qualities.

The procedure developed to control locusts by predation with guinea fowl is not only an achievement in the history of plant health but also demonstrates the extraordinary usefulness of this bird in enriching the biodiversity of species of agricultural interest in sustainable agriculture.

Apart from this achievement, the studies carried out to develop this procedure for locust predation also served to learn a large number of techniques for managing the guinea fowl—artificial insemination, breeding and rearing of guinea fowl, feeding programme, health, construction, etc. The extraordinary hardiness of this species had been proven, and it seemed advisable to test extensive rearing of these birds in which the guinea fowl were kept in the open air during the entire rearing period, for the study of which it would be necessary to compare this system with an intensive system. A project financed by the Regional Government of Extremadura (Spain) and the EU (ERDF) was carried out for this purpose and concluded that, with the exception of the coldest months of the year, extensive rearing of guinea fowl is possible and that the carcasses produced are very rich in heart-healthy fatty acids (mono- and polyunsaturated) and have a higher organoleptic quality than those obtained in the intensive system. In parallel to the previous project, another project was carried out to determine the shelf life of guinea fowl carcasses subjected to different treatments and packaging systems, thanks to which it has been possible to establish a procedure for the preservation of carcasses.

During all this time of study and experience, renowned professionals from the restaurant and catering sector were invited to check the quality of the meat being produced, and they were astonished. The carcasses they were given to compare with those they buy from France—intensively produced—were, they said, excellent.

Nowadays, it is difficult to find a species like the guinea fowl that can be used with such versatility in agriculture: it can be used as a lobster predator, and its meat is excellent, and it can be farmed extensively, in the open air, producing eggs and carcasses that surpass the quality of those on the market from other countries. At the same time and given that its flight characteristics are similar to other birds used in hunting activities (partridge and pheasant), it can also be used for this purpose, as is the case in Italy and some countries in America (Mexico, Cuba, etc.) and Africa (Kenya, Mozambique and many other countries).

This species is of utmost importance to agriculture in France and Italy. The market value of its carcasses exceeds half a billion euros/year, as well as its contribution to the biodiversity of agriculture and to the richness of gastronomic tourism, highlighting the economic potential of its livestock.

This book contains the technology necessary to integrate this species into the sustainable agriculture that needs to be implemented in all rural areas of the planet.

There is one final consideration that is worth reflecting on in these times of crisis, times when we are obliged to be diligent in the design of the new sustainable agriculture. The technology contained in this book on guinea fowl may be very valuable, but we cannot forget that in Hispania (the name of Spain in the Roman Empire), two thousand years ago, these birds were already being exploited. How many plant and animal species, and how many ancient techniques that are stored in the attic of agriculture in the depopulated rural areas of the planet, could be rescued and adapted to sustainable agriculture worldwide?

Guadajira, Badajoz, Spain — José del Moral de la Vega
Badajoz, Spain — Jerónimo del Moral Martínez

Acknowledgements

To Juan María Vázquez García, Antonio Vélez, Francisco Quintana Gragera, Julio Domínguez Merino, Francisco Martos Ortiz and Daniel Del Castillo Corvillo, who facilitated the research on the guinea fowl.

To Gregorio Hernández Silva, who provided us with the first bird trunk.

To the scholarship holders Dolores Corrales Moreno and Nuria Martínez Doncel, and to the laboratory and field assistants Víctor Andújar Romero, Modesto Mayoral Rodríguez and Fátima Pérez Rojas, for their participation in various project tasks.

To Mariano Cancho Alfonso, Mª José Contreras de Vera, Juan Manual Cumplido García, Antonio García García, Paloma Ledesma Jurado, Fermín López Gallego and Antonio Prieto Andreu, researchers and technicians who have helped us to solve numerous problems that have arisen during the research.

To Professor Enrique Campillo Álvarez, for his assessment of the nutritional analysis of various guinea fowl foods.

To the agricultural entrepreneurs and cooperative directors, Javier Donoso Caro, Manuel Rodríguez Corrales and Cristóbal Álvarez del Bayo (ACOREX); Juan Cabanillas Carpio, Manuel Donoso Donoso, Florencio Villarubias, Miguel Herrojo González, Francisco J. González Pizarro; José A. Carrizosa Molina and José Miguel Rivero Gómez (COEX), Juan Carlos Pozo (OVISO) and Antonio Muñoz (AMF), who have collaborated with their farms, companies and advice in the development of the projects on the bird.

To Julio Domínguez Merino, former mayor of Azuaga (Badajoz), for his help in building a farm for raising guinea fowl.

To the professionals, directors and entrepreneurs of the gastronomic tourism sector: Carmen Comino García (Director of the Parador Nacional in Mérida), Juan Sanguino Gallardo (Chef of the Parador Nacional in Mérida), Javier García Guerra (Chef of the Lugaris restaurant, Badajoz), David Chapela (Chef of the Zurbarán restaurant, Badajoz), Nicasio Durán (Aldebarán Restaurant, Badajoz), Agustín Fernández (Chef of the La Alacena restaurant, Badajoz), Mercedes Ferreras (Teacher at the San Fernando Catering High School, Badajoz), Iván Hernández Rivera (Chef of the Gran Casino de Extremadura), Antonio López Archilla (Director of the Parador Nacional in Zafra), Pedro Camacho Merino (Chef of the Parador Nacional in Zafra), Antonio Murillo Castillo (Restaurante La Dehesa, Azuaga), Julia Navarro Machín (Director of the Parador Nacional in Calahorra), Ernesto

Osoro Gorrotxategi and Alonso Pantoja Lagar (Teachers at the IES de Hostelería San Fernando, Badajoz), Juan Pedro Plaza Caravantes (Director of the Tourist Board of the Badajoz Provincial Council), Félix Retamar Martín (Former director of the Hotel Zurbarán, Badajoz), Miguel Angel Vélez (Chef of the Los Monjes restaurant, Badajoz), Pedro Sánchez Montero (Bodegas Las Granadas Coronadas, Coria).

To illustrators Diego Del Moral Martínez and José Del Moral Martínez for providing the excellent drawings that illustrate some of the chapters in this book.

To the media, Periódico HOY and El Periódico Extremadura; radio stations COPE, SER and Onda Cero; television stations, la 2 de TVE, Extremadura TV and Antena 3. To journalists Luis Expósito, Judit Figueiredo, Silvia García García, Julián Leal, Guadalupe Leitón Caro, Paco Moreno, María Ortiz Alvarado, Guadalupe Peromingo, Julia Sancho and Antonio Tinoco Ardila.

To Teresa Hernández Valiente, for the adaptation of the publications into English; to Sophie Fernández Walters, for proofreading the English translation of this book; and to Professor Angélica Beatriz Baca Pérez, for the final reading of the text and the correction of the numerous errors it contained.

To the Spanish Plant Health Association (AESAVE) for recognising the contribution of this work to improving global plant health.

This book is the result of an initial concern awakened in the social interest gatherings that spontaneously took place in 1979 in the Faculty of Sciences of the University of Extremadura, and in which then university students Ramón de Arcos Nieto Guerrero, José González Crespo, Pedro Gómez Hernández and José del Moral de la Vega—today university professors or researchers—discussed with the Professor of Ecology, José Carlos Escudero García, how the University of Extremadura could contribute to solving various problems in agriculture and the rural world.

The techniques presented here have been obtained from the following projects: "Study to assess the effectiveness of locust (*Dociostaurus maroccanus* Thunb.) pest control by means of bird predation", "Comparison of two guinea fowl (*Numida meleagris* L.) production systems to determine the quality of the carcasses produced and their economic viability in Extremadura", "Transfer to the company AMETUGA SA. of a technology obtained by the Junta de Extremadura for the multiplication and breeding of guinea fowl in La Campiña" and "Study of the useful life of guinea fowl carcasses subjected to different treatments and packaging systems. Organoleptic quality". These projects have been financed by INIA, the EU (ERDF) and the Regional Government of Extremadura.

And finally, thanks to Complementary Action N°: AC2009-00050-00-00 financed by INIA, the Regional Government of Extremadura and EU ERDF funds.

Contents

Editors and Contributors

About the Editors

José del Moral de la Vega Emeritus Head of the Department of Plant Pathology of CICYTEX, (Scientific and Technological Research Centre of Extremadura), CICYTEX-Finca La Orden, Guadajira, Badajoz, Spain

Jerónimo del Moral Martínez Food Quality and Microbiology Research Group (AGA015) of the Extremadura Science, Technology and Innovation System, (Badajoz), Spain

Contributors

David Tejerina Barrado Meat Quality Area at CICYTEX (Scientific and Technological Research Centre of Extremadura), Badajoz, Spain

Modesto Senero Fernández CICYTEX (Scientific and Technological Research Centre of Extremadura), Guadajira, Badajoz, Spain

Susana García-Torres Meat Quality Area at CICYTEX (Scientific and Technological Research Centre of Extremadura), Badajoz, Spain

Ángel Mejías Guisado CICYTEX (Scientific and Technological Research Centre of Extremadura), CICYTEX-Finca La Orden, Guadajira, Badajoz, Spain

Cesáreo Sánchez López CASTRUM ERAT Farm, Castuera, Extremadura, Spain

María Jesús Martín Mateos Meat Quality Area at CICYTEX (Scientific and Technological Research Centre of Extremadura), Badajoz, Spain

Javier Jiménez Medina FULGENCIO SPA, S.L., Granada, Spain

Víctor Javier Parralejo Moreno Locust Biological Control Working Group, Department of Plant Pathology, CICYTEX (Scientific and Technological Research Centre of Extremadura), Guadajira, Badajoz, Spain

María Montaña López Parra Meat Quality Area at CICYTEX (Scientific and Technological Research Centre of Extremadura), Badajoz, Spain

Francisco Pérez Rojas CICYTEX (Scientific and Technological Research Centre of Extremadura), Guadajira, Badajoz, Spain

Pedro Emilio Rosado Salgado Locust Biological Control Working Group, Department of Plant Pathology, CICYTEX (Scientific and Technological Research Centre of Extremadura), Guadajira, Badajoz, Spain

Carmen Sanabria Tienza CICYTEX (Scientific and Technological Research Centre of Extremadura), Badajoz, Spain

Abbreviations

CFUs	Colony Forming Units
CRD	Chronic Respiratory Disease
FAO	Food and Agriculture Organization of the United Nations
IM	Individual Motility
INIA	(Instituto Nacional de Investigaciones Agrarias) Spanish National Institute for Agricultural Research
INRA	(Institut National de la Recherche Agronomique) French National Institute for Agricultural Research
ME	Metabolisable Energy
Mi.M.M	Microscopic Mass Motility
MUFA	Total Monounsaturated Fatty Acids
PA	Crude Protein
PUFA	Total Polyunsaturated Fatty Acids
SFA	Total Saturated Fatty Acids
SHS	Swollen Head Syndrome
TRT	Turkey Rhinotracheitis
WPSA	World's Poultry Science Association

Introduction

1

José del Moral de la Vega
and Jerónimo del Moral Martínez

Abstract

The discovery of *Numida meleagris* (Linnaeus, 1758), known as the guinea fowl, was made by the Egyptians (eighteenth century BC) in their explorations at the source of the Nile, and its use spread throughout the Mediterranean, thanks to the Greeks and Romans, whose culinary preparations are mentioned in the first century—*Re coquinaria libri decem*. Portuguese and Spanish explorers of the Modern Age brought them from Africa to America, and since then, there are numerous reports of their domestication and use in many countries of those continents. The excellence of its meat, considered superior to that of the pheasant or partridge, has led to its exploitation as livestock in various countries, especially in France, the world's leading producer. Knowledge of these characteristics, together with the observation of its behaviour as an insect predator in the farmhouses of Southern Spain, where it has existed since time immemorial, has led to the development of various research projects which have demonstrated its potential to be reared on extensive livestock farms, as well as its usefulness as a predator of locust pests. Organoleptic analyses carried out on the carcasses of these birds have revealed that their quality is significantly higher than those reared in intensive livestock systems.

J. del Moral de la Vega
Emeritus Head of the Department of Plant Pathology of CICYTEX, (Scientific and Technological Research Centre of Extremadura), CICYTEX-Finca La Orden, Guadajira, Badajoz, Spain

J. del Moral Martínez (✉)
Food Quality and Microbiology Research Group (AGA015) of the Extremadura Science, Technology and Innovation System, (Badajoz), Spain

J. del Moral de la Vega, J. del Moral Martínez (eds.), *The Guinea Fowl (Numida meleagris)*, https://doi.org/10.1007/978-3-032-11654-3_1

Keywords

Guinea fowl · Extensive production · Quality meat · Mediterranean biome · Pest control · Sustainable agriculture

References to This Bird in Historical Bibliographic Collections

Until the end of the last century, when the countryside was full of people and chores, the farmhouses and country houses of Spain were often full of chickens, turkeys and other birds similar to these (Fig. 1.1), but screaming and flighty, with a rather small head with a protuberance or horn, and with feathers covered with grey speckles, which is why Spanish farmers called them *pintadas*, a name that alternates with *gallina de Guinea*, a probable translation into Spanish of the one that appears in many French publications. The truth is that although the farmers were not well acquainted with their breeding, and their consumption was not widespread, their presence was somewhat extravagant in Spanish farming culture, and their origin was not even known.

We must remember that our civilisation is like a treasure chest from which we can always find what we are looking for.

Fig. 1.1 When farmhouses and country houses were a vibrant of the Spanish rural world, their corrals used to be full of birds, among which the guinea fowls' once stood out for their cries and savagery

Ancient references to the speckled bird are not frequent in the world's bibliographic collections of science or technology, but if we rummage through the collections of Greek Mythology, we may be surprised by a very interesting story about these birds:

> Meleager was a Greek prince, son of Oeneus (king of Calydon) and Althea (lover of Dionysus). When he was born, the Fates announced to Althea that Meleager's life would end when a log that was burning at the time died out, which is why she set it aside and kept it in a chest.
>
> One day, Oeneus was offering sacrifices of thanksgiving to the gods and forgot to invoke Artemis (daughter of Zeus, and goddess of hunting and chastity), so, angry, she sent a rabid boar to Calydon that destroyed and killed everything in its path. Many heroes joined forces to kill it, including Meleager's uncles (the Thespians) and a beautiful woman, Atalanta. Together they killed the boar, although it was Meleager who delivered the final blow, and that is why he would be the one to receive the trophy. However, the hero fell in love with the beautiful Atalanta and gave her the boar, which caused offence to the Thespians who claimed it as their right, creating a conflict that led to a battle between them and Meleager, in which Meleager killed his uncles.
>
> When Althea heard that her brothers were dead, she became enraged and began to throw everything she had at hand into the fire, including the log she had kept for so long, whereupon Meleager—as the Fates had foretold—died. Althea, full of remorse, hanged herself, and Meleager's sisters, heartbroken, wept over their brother's grave. The goddess Artemis, moved by their sorrow, transformed them into birds, giving them grey feathers speckled with white, like the tears they still shed for their brother. (García 2023)

This is the myth that explains the origin of the guinea fowl (Fig. 1.2) and which justifies Linnaeus calling this species *meleagris* and the genus *Numida*, a name that refers to the Berber region of North Africa (*Numidia*) where these birds were abundant in ancient times and from where, at first, it was believed that they had come from.

The Greeks must have been introduced to these birds by the Egyptians, who had integrated them into their cuisine, although the first news of their use comes from the Romans—in the *De re coquinaria libri decem*, by Marcus Gavius Apicius, there is a recipe made with guinea fowl meat, which he calls "Guinea hen". It is very likely that the Egyptians knew of these birds through explorations along the Nile, as they are thought to originate near the source of this African river (Toynbee 1973).

Some authors claim that guinea fowl were domesticated in Egypt and Greece around 1745 BC and 400 BC, respectively, and genetic analysis places its domestication in original populations in West Africa (Vignal et al. 2019). It is now believed that the guinea fowl bred for livestock were brought to Europe by Portuguese and Spanish navigators (Vasco de Gama, Bartolomé Díaz), who also transported them to the Americas. In fact, in Puerto Rico, it is known that the Guinea fowl were brought there, in 1549, by Don Diego Lorenzo, a canon who was from Cape Verde (Taveras and Mattei 2023) (Fig. 1.3), although Columela, who is considered the first Spanish agronomist (first century BC), already mentions them, calling them African or Numidian fowl (Sutherland 2024).

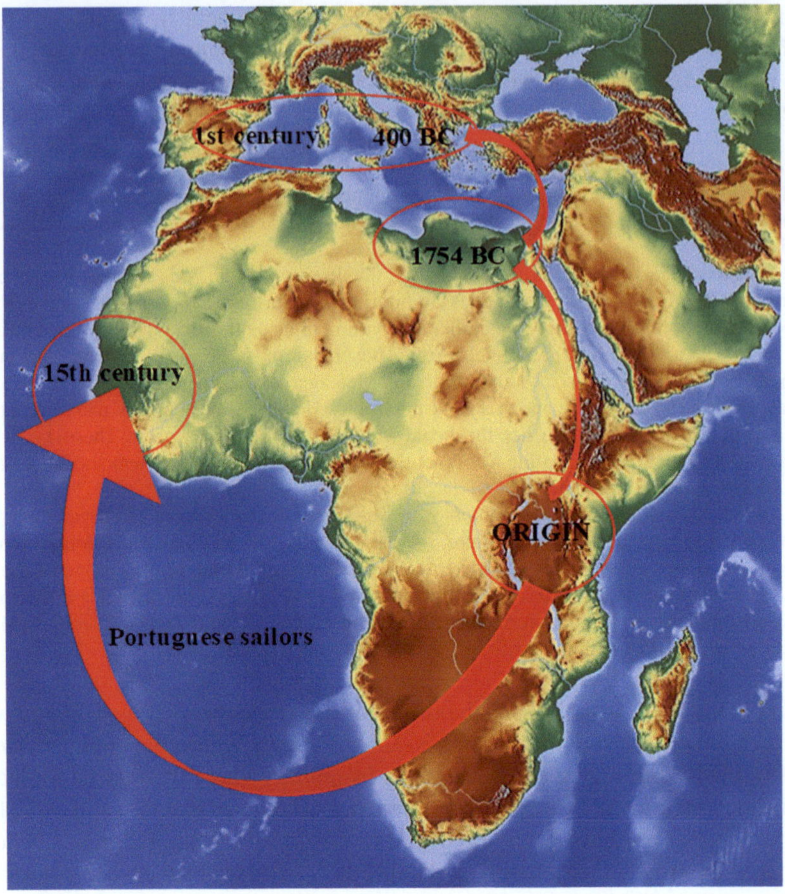

Fig. 1.2 There is considerable evidence that guinea fowl (*N. meleagris*) originated in a place near the source of the Nile. From there, accompanying the development of civilisation, it passed to Egypt, Greece, Rome, etc. (Image created by us using images in the public domain from the digital platform https://pixabay.com/)

Production and Consumption

Guinea fowl farming mainly takes place in Europe, but it is also practised in various countries in other continents, such as the United States, Canada, Brazil, Madagascar, China and Russia. European production is confined to a few European countries (Fig. 1.4), although France is the world's leading producer (Araújo et al. 2023).

Guinea fowl farming in France takes place on over 700 farms, mostly in Pays de la Loire and Bretagne but also in Rhône-Alpes and Poitou-Charentes. Selected by seven main companies (Galor, Le Sayec, Beghin, Goubin, ISA, Grimaud, Hendrix) with a hatching capacity of nearly 50 million eggs, guinea fowl are reared in three

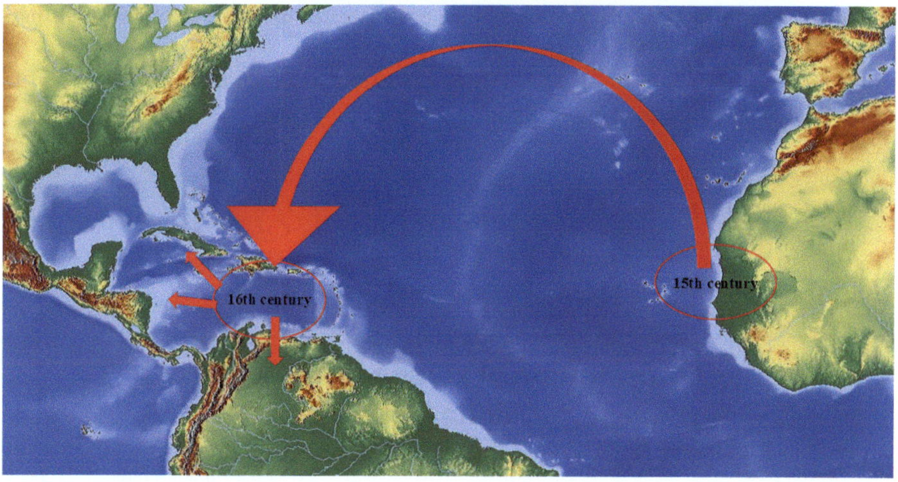

Fig. 1.3 There are documents proving that guinea fowl (*N. meleagris*) were transported in the fifteenth century by Portuguese sailors from the eastern coasts of Africa to the western coasts, and from there, accompanying missionaries, slaves and merchants, they reached the West Indies (sixteenth century), from where they spread to the whole American continent. (Image created by us using images in the public domain from the digital platform https://pixabay.com/)

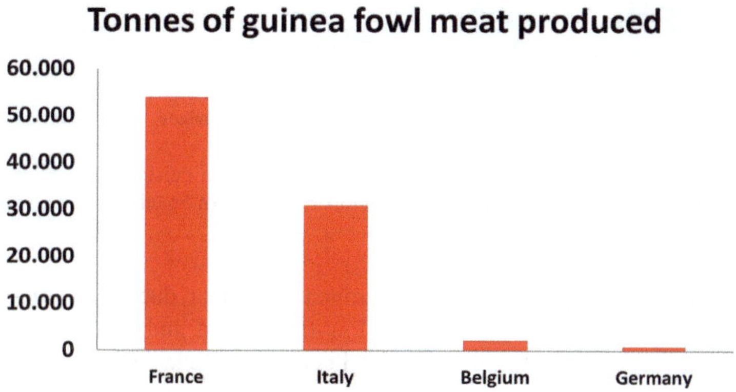

Fig. 1.4 Guinea fowl meat production in the leading European producing countries (Cook 2020)

distinct ways: Standard, Labelled—Labels—and Protected Designation of Origin (PDO) *Pintadeau de la Drôme* (Le Coz 1992) (Table 1.1).

The way of producing guinea fowl in the European Union, like that of any other livestock, tends to be standardised to ensure homogeneity, quality and safety, a trend that is driven by the price consumers are prepared to pay. In France, for example, the price per kg of "Label" poultry is usually 1.6 times higher than that of the standard version. This production appeal has resulted in France currently having 14 labels, differentiated according to French administrative rules regarding traceability, farming methods, feeding standards, prophylaxis, age and slaughter conditions.

Table 1.1 Most important characteristics of the poultry reared in France according to the three types of guinea fowl produced

	Standard	Designation of origin	Label
Density of birds (individuals/m^2)	16.2	15	13
Parks	No	Yes	Yes
Slaughter age (days)	82	87	94
Slaughter weight (kg)	1.55	1.75	1.78
Conversion rate (kg feed/kg livestock per animal)	2.95	3.25	3.7
Poultry production (in millions)	37	4	5

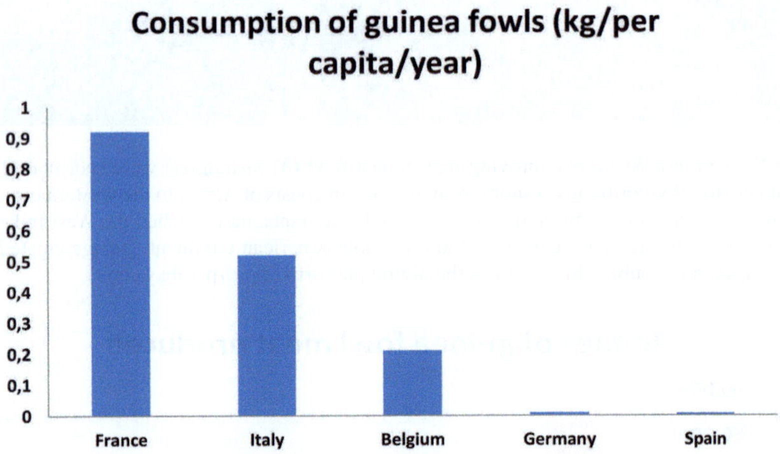

Fig. 1.5 The consumption of guinea fowl in French and Italian cuisines—the world's leading gastronomic tourism powers—is extraordinary compared to that of Spanish cuisine, where guinea fowl meat is only present in top-level restaurants

The nutritional and hygienic value of guinea fowl meat, due to its composition in proximate principles, mainly due to its low-fat content and quality, is excellent. However, the meat is most highly regarded in culinary circles. Dishes prepared with this meat are held in high prestige in French and Italian haute cuisine, even surpassing those of other game meats such as pheasant or partridge, which is why consumption figures in these countries are considerable (Araújo et al. 2023) (Fig. 1.5).

Potential for Livestock Farming and Other Uses of Guinea Fowl in the Mediterranean Biome

France is the world's leading guinea fowl producer, and most of its production, except for some types of "Label", is carried out intensively, with the birds kept all day or most of the day in enclosed facilities at a density of around 15 birds per m^2.

This system of production has the advantage of a high conversion rate (kilograms of feed needed for 1 kg weight gain) due to the birds not being able to move, although this also increases the risk of disease which requires strict preventative measures. Nonetheless, the lack of exercise causes a reduction in the quality of the meat.

Most of the Mediterranean biome is characterised by low rainfall, dry soil, sunny winters and not many days with temperatures below 0 °C.

These characteristics suggested that a free-range guinea fowl farming system could easily be developed in these areas, where the birds would be outdoors for the duration of their rearing, starting from the sixth week after hatching. This speculation resulted in studies in Spain comparing an extensive farming model, with the birds living outdoors, with another in which the guinea fowl were always kept indoors (Rosado et al. 2011). These studies served to demonstrate the adaptability of this bird to outdoor living, showing that, in Southern Spain, extensive livestock farming is possible during most of the year, with the birds only being housed during the colder months (December, January and February).

The viability of the extensive guinea fowl system in the Mediterranean biome is complemented by two features that add further advantages to this system.

The labour costs required by the intensive system are reduced, and above all, the carcasses obtained are of significantly higher quality than those produced on farms where the birds are confined for their entire lives.

In organoleptic studies carried out on the parameters of colour, flavour, tenderness, juiciness and total evaluation, significant differences have been obtained between extensively reared and intensively reared guinea fowl meat, providing unquestionable positive results on the quality of dishes prepared with extensively reared guinea fowl meat (Rosado et al. 2011) (Fig. 1.6).

It is clear that although this extensive guinea fowl farming method yields a lower carcass weight than the intensive system, the quality of the meat obtained is far superior, as is the welfare of the animals reared in "quasi" freedom. Both aspects are highly valued by consumers today and in line with the guidelines for future livestock farming systems in the European Union.

Another aspect to consider in the usefulness of this bird is its ability to prey on insects, a characteristic well known to Spanish farmers.

In order to find an alternative biological procedure to the use of synthetic insecticides to control locust (*Dociostaurus maroccanus* Thunb.) pests that are endemic in South-western Spain, studies have been carried out to assess the predatory capacity of these birds (Del Moral et al. 2006). These studies have demonstrated their effectiveness—the predator eats the locust, and humans eat the guinea fowl, transforming an infestation into an added value. The quality of the meat thus obtained, assessed using a protocol recommended by the "Working group of WPSA for sensory analysis of broilers", has shown that when the guinea fowl ingest insects, the meat produced significantly improves the organoleptic values analysed (flavour, juiciness, tenderness, fat and overall value), with it being rated as exquisite by the six expert tasters who analysed it (Rosado et al. 2011). It is clear that the use of guinea fowl as a predator of locust pests increases the potential of this bird.

Fig. 1.6 Guinea fowl meat reared extensively has been classified, according to rigorous tasting procedures, as much better in comparison to birds reared in confined poultry houses. This assessment has coincided with customer surveys of guinea fowl dishes prepared in some of the best restaurants. Guinea fowl recipe by Javier García Guerra, chef at Lugaris, Badajoz (Extremadura, Spain) (Confit guinea fowl breast with caramelised onions)

Another use for this bird is hunting, a sporting activity that generates wealth in the rural world.

In various African countries, such as Kenya, Mozambique, Tanzania, Guinea Bissau and South Africa, guinea fowl that live in total freedom are hunted like any other wild animal. Conversely, in parts of the Americas (Cuba, Mexico, Puerto Rico) where this bird was introduced and has become fully acclimatised, it is hunted in preserves using various procedures including leaping and controlled releases. The management and improvement of this type of hunting represent a powerful source of income for the sustainability of rural communities (Anderson and Holmes 2022; Aynalem et al. 2023a, b; Van Niekerk et al. 2024).

The Guinea Fowl in the New Sustainable Agriculture Promoted by the FAO

Global agriculture is currently in crisis because a large part of this sector is being taken out of developed countries and transferred to underdeveloped or developing countries. While this shift does not seem to cause major problems at a national level, it does have a significant impact at a regional level, especially when it affects smaller population centres. This does not seem to cause major problems at the state level,

Fig. 1.7 In order to maintain the rural population, a new agriculture is needed, for which sustainable models must be designed, closely linked to the industrial and service sector, mainly gastronomic tourism, in which extensive guinea fowl farming can play an important role

but it does at the regional level and even more so when it is concentrated in smaller population centres. What is the future of purely rural areas (Fig. 1.7), which are much less developed than industrial and service-oriented areas?

In order to avoid a social catastrophe there, it is necessary to develop alternative agricultural models that are sustainable and work in tandem with the service sector (rural and gastronomic tourism).

It is clear that today's agriculture, if it is to be sustainable, must be environmentally friendly, competitive and socially useful.

One of the ways outlined for the implementation of sustainable models of agriculture is the diversification of species for their use: first, by harmoniously introducing those that contribute to the balance of agroecosystems and, second, by promoting the transition from intensive systems to sustainable models, usually through extensive systems (Canton 2021; Liu 2023; Salisu et al. 2024; Pakeerathan 2025).

In addition to their adaptability to the typical Mediterranean environment, studies carried out in South-west Spain on these birds in an extensive regime have shown their economic viability (Rosado et al. 2011). Their infrastructure and management needs are low, which is why extensive guinea fowl production could complement the work of agricultural or livestock farms in Southern Spain and increase their income, thereby contributing to the sustainability of the rural world.

The guinea fowl consumed in Europe is mainly produced under intensive systems, but the market shows a trend toward quality meat produced under organic or "quasi-organic" systems, a trend that increases each time a new scandal breaks about anomalies in the exploitation of animals for human consumption. However, extensive livestock farming systems are impossible to implement, or are very costly, when they are developed in countries or regions in Central and Northern Europe

(Germany, Sweden, Finland) or in areas with a high level of income (France, Italy or Belgium). The southernmost areas of Europe on the other hand are ideal for their implementation, as are regions similar in biome on other continents, mainly the Americas and Africa (Oguche et al. 2025).

In summary, there is a complete and proven extensive model for breeding and rearing guinea fowl in South-west Spain; the quality of the carcasses produced in South-western Spain using the extensive system has been rated much higher than that of meat produced using an intensive system, and the system can be extrapolated to other regions of the planet with similar biomes.

If all these conditions continue to occur in the immediate future, as is to be expected, it is evident that the development of extensive guinea fowl farming in all those areas of the planet that share the Mediterranean biome can play an important role in the sustainability of their agriculture and in the consequent maintenance of the rural world that lives from it.

Competing Interests The authors have no conflicts of interest to declare that are relevant to the content of this chapter.

References

Anderson G, Holmes C (2022) Galliformes: species, varieties, and behavior. In: Gamebird medicine and management. Wiley, pp 13–35

Araújo ICS, Guato Guamán C, Sousa LS, Santos HJB, Lopes TSB, Costa BTA, Lara LJC (2023) Guinea fowl production in the world. Worlds Poult Sci J 79(2):379–390

Aynalem Y, Bekele A, Bogale BA (2023a) Breeding biology of the helmeted guineafowl (*Numida meleagris meleagris*, Linnaeus 1758) in Western Ethiopia. Ornithol Res 31(3):173–181

Aynalem Y, Bekele A, Bogale BA (2023b) Population size, habitat use and foraging habits of wild helmeted Guineafowl *Numida meleagris* in SouthWestern Ethiopia. Ostrich 94(3):228–233

Canton H (2021) Food and agriculture organization of the United Nations-FAO. In: The Europa directory of international organizations 2021. Routledge, pp 297–305

Cook E (ed) (2020) Agriculture, forestry and fishery statistics: 2020th edition. Publications Office of the European Union

Del Moral J, Mejías A, Jiménez J, Pérez-Rojas F, Senero M (2006) Control de las plagas de langosta (*Dociostaurus maroccanus* Thunb.) con el ave "pintada" (*Numida meleagris* L.), un procedimiento ecológico alternativo o complementario a la aplicación de insecticidas. Acta Cient Tecnol (Spain) 11:9–13

García JT (2023) El lamento fúnebre de las meleágridas en los sarcófagos romanos: una propuesta de interpretación gestual y fisiognómica. Codex Aquilar Cuad Investig Monast Santa María Real 39:13–34

Le Coz DJ (1992) L'élevage de la pintade. Edition point vétérinaire/Maison Alfort. 252p

Liu S (2023) Towards a sustainable agriculture: achievements and challenges of sustainable development goal indicator 2.4. 1. Global Food Secur 37:100694

Oguche M, Kariuki J, Birner R, Chagunda M (2025) Is there unrecognized potential in neglected livestock species in sub-Saharan Africa? A systematic review of four selected species: unrecognized potential in neglected livestock species in Africa. Food Secur 17(1):161–183

Pakeerathan K (2025) A global overview and the fundamentals of sustainable agriculture. In: Hyperautomation in precision agriculture. Elsevier, pp 3–13

Rosado PE, Pérez F, Martínez N, Senero M, Del Moral J (2011) Producción extensiva de" pintadas" o gallinas de Guinea en Extremadura. Sel Avícolas 53(3):61–66

Salisu MA, Ismail F, Bamiro NB, Luqman H (2024) Sustainable agriculture for food safety, security, and sufficiency. In: Agripreneurship and the dynamic agribusiness value chain. Springer Nature Singapore, Singapore, pp 29–60
Sutherland A (2024) The cock in antiquity: revered, venerated, deified. In: Cockfighting in Britain from antiquity to the enlightenment. Springer Nature Switzerland, Cham, pp 15–39
Taveras JA, Mattei L (2023) San Germán en Guayanilla: asentamiento, instituciones y vida cotidiana, 1556–1572. Palibrio
Toynbee JMC (1973) Animals in Roman life and art. Cornell University Press, Ithaca
Van Niekerk JH, Forcina G, Megía-Palma R (2024) Grouping behaviour and anti-predator responses in the helmeted Guineafowl *Numida meleagris*. Birds 5(4):685–702
Vignal A, Boitard S, Thébault N, Dayo GK, Yapi-Gnaore V, Youssao Abdou Karim I, Berthouly-Salazar C, Pálinkás-Bodzsár N, Guémené D, Thibaud-Nissen F, Warren WC, Tixier-Boichard M, Rognon X (2019) A Guinea fowl genome assembly provides new evidence on evolution following domestication and selection in galliformes. Mol Ecol Resour 19(4):997–1014

Biological Characteristics

2

Víctor Javier Parralejo Moreno

Abstract

The guinea fowl *Numida meleagris* (Linnaeus, 1758) is a bird species belonging to the Galliformes order, family Numididae, subfamily Numidinae, originating in Africa, of which there are currently about ten known subspecies. Domesticated specimens differ very little from wild specimens, weighing more than 2 kg and laying more than 150 eggs per year. Their eggs are slightly smaller and denser and have thicker shells than those of the domestic chicken (*Gallus gallus* L.), which makes them easier to transport. A gregarious but monogamous bird, it is highly resistant to inclement weather, except for extremely cold temperatures.

Keywords

Guinea fowl · *Numida meleagris* · Characteristics · Behaviour

Systematics

The guinea fowl is a bird species originally classified in the order Gallinacea, suborder Alectoropoda, family Phasianidae, subfamily Numididae. From this classification, it is clear that guinea fowl are related to pheasants, but many authors have argued that the Numididae should not be considered as a subfamily but rather as a family completely independent of the Phasianidae. Therefore, the scientific classification of this species is now determined as follows, and its binomial name is *Numida meleagris* (Linnaeus, 1758).

V. J. P. Moreno (✉)
Locust Biological Control Working Group, Department of Plant Pathology, CICYTEX (Scientific and Technological Research Centre of Extremadura), Guadajira, Badajoz, Spain

J. del Moral de la Vega, J. del Moral Martínez (eds.), *The Guinea Fowl (Numida meleagris)*, https://doi.org/10.1007/978-3-032-11654-3_2

Kingdom: Animalia
Phylum: Chordata
Class: Birds
Order: Galliformes
Family: Numididae
Genus: *Numida* (Linnaeus, 1766)
Species: *N. meleagris*

Within the family Numididae, there are five genera characterised by head ornaments: four living and one fossil (*Telecrex* Wetmore 1934). Among them, the genus *Agelastes* (Bonaparte, 1850), which mainly inhabits tropical Africa, includes *Agelastes meleagrides* (Bonaparte, 1850) and *Agelastes niger* (Cassin, 1857), both of which are completely "acrested". The royal guinea fowl, *Acryllium vulturinum* (Hardwicke, 1834), also commonly known as *vulturina*, has only a few short feathers as its cephalic ornament. It is the largest, most beautiful, and gentlest guinea fowl but also the most difficult to adapt to cooler climates and is mainly found in Somalia, Tanganyika and Zanzibar. The plumed guinea fowl, *Guttera plumifera* (Cassin, 1857) and *Guttera pucherani* (Hartlaub, 1861), are characterised by their feathered head, while the throat and chin are devoid of feathers. They are mainly found in tropical African countries but have also been found further south on the continent. Finally, the helmeted guinea fowl are those of the genus *Numida*, which, as their name suggests, are characterised by the presence of a horny cap or crest on the top of the head. They also bear caruncles, the colour of which allows the distinction between two groups: those with blue caruncles and those with red caruncles, the red caruncled being the most adaptable and the origin of the domestic guinea fowl. There are ten subspecies of *Numida meleagris* depending on their origin:

1. *Numida meleagris sabyi* Hartert 1919 → north-western Morocco
2. *Numida meleagris galeatus* Pallas 1767 → from West Africa to southern Chad, central Zaire and northern Angola
3. *Numida meleagris meleagris* (Linnaeus 1758) → from eastern Chad to Ethiopia, northern Zaire, Uganda and northern Kenya
4. *Numida meleagris somaliensis* Neumann 1899 → northeastern Ethiopia and Somalia
5. *Numida meleagris reichenowi* Ogilvie-Grant 1894 → Kenya and central Tanzania
6. *Numida meleagris mitratus* (Pallas 1764) → from Tanzania to eastern Mozambique, Zambia and northern Botswana
7. *Numida meleagris papillosus* Reichenow 1894 → Namibia, Botswana, Cape Verde and Senegal
8. *Numida meleagris marungensis* Schalow 1884 → from the southern Congo Basin to western Angola and Zambia
9. *Numida meleagris damarensis* → from southern Angola to Botswana and Namibia

10. *Numida meleagris coronatus* Gurney Sr 1868 → eastern South Africa (Belshaw 1985; GBIF 2023; Clements et al. 2024; Natural History Museum 2024)

Characteristics of the Bird

The common guinea fowl (Fig. 2.1) is a wild, domesticated bird since ancient times. Yet despite this taming, it still exists on the border between the wild and the world of the farm. This is evident when wild guinea fowl eggs are taken and incubated with a common hen; the chicks hatched from these eggs display the same behaviour as the chicks of the hens on the farm. Conversely, adult farm-raised guinea fowl, when released into the wild, adapt perfectly and reproduce normally. This occurs not only when they are returned to their natural environment but has also been achieved in colder regions such as France, where tests have been carried out for hunting purposes.

Fig. 2.1 Although guinea fowl resemble chickens, hence their name, their bodies are more slender. (Graphite drawing by Diego Del Moral Martínez)

It should be noted, however, that there are differences between the wild guinea fowl found in West African countries and the domestic guinea fowl seen on many farms in Europe. The domestic guinea fowl is characterised by having a slightly higher live weight than that of the wild guinea fowl. In addition to laying larger eggs; the wattles of the domestic guinea fowl are larger, and their tarsi are not as pigmented. These small differences are due to the genetic selection that this domestic guinea fowl has undergone; the main characteristics of which are described below (Ajenjo 1964; Le Coz-Douin 1992).

Morphology

The guinea fowl is referred to by farmers as the "guinea hen" because of its morphological resemblance (Fig. 2.2) to the domestic chicken (*Gallus gallus* L.), although its appearance varies depending on whether the bird is bent toward the ground or standing up. In the latter case, its figure is more stylised, with an ovoid body, a slender conical neck and a tail inclined downwards, with a total length of 53–63 cm and a wing chord of 247–276 mm.

The head is small, rather short but broad, is featherless and covered with a slightly bluish-white skin with small black pigmented patches. The blue colour of the skin is accentuated around the eyes and below the neck; in this area, this colour

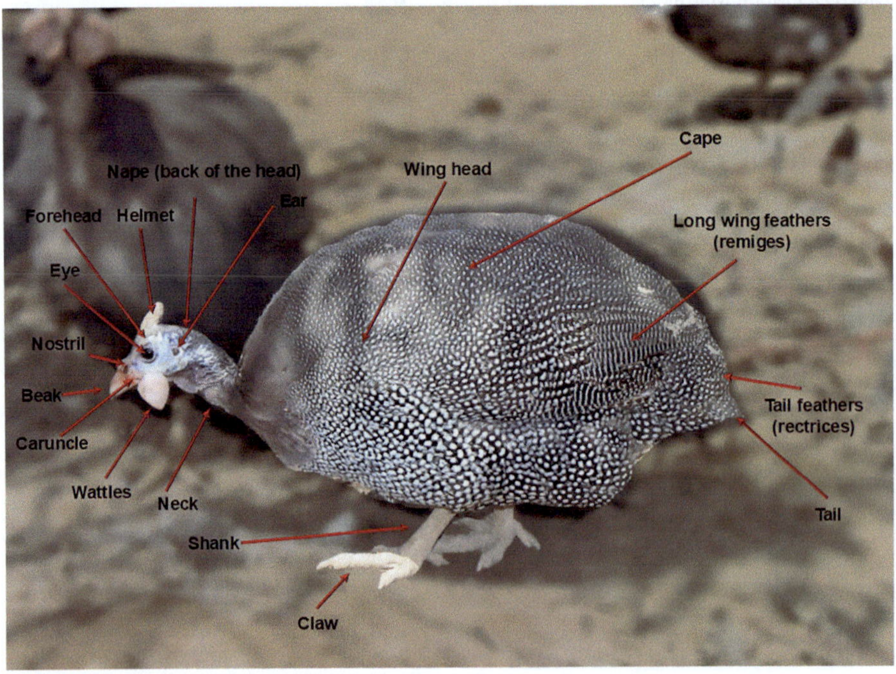

Fig. 2.2 The most notable morphological characteristics of the body of painted birds

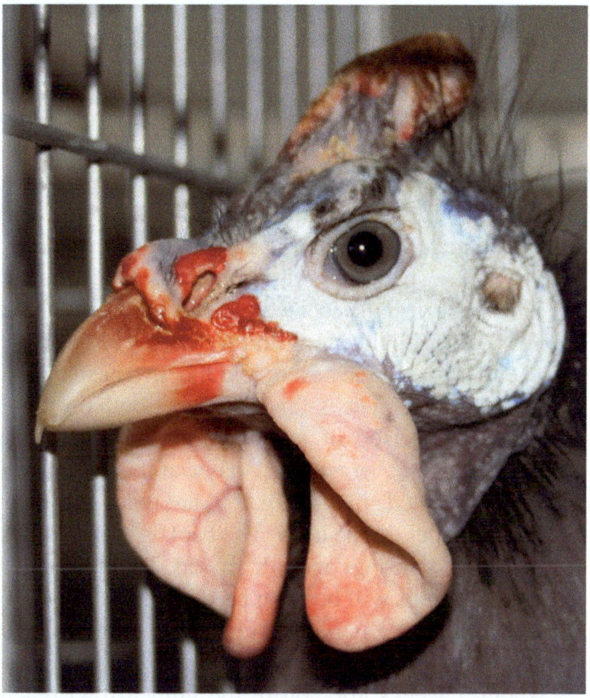

Fig. 2.3 The horny crest (helmet) crowning the bird's body, the caruncle above the beak and the hanging wattles characterise the head of guinea fowl

stands out, giving the appearance similar to a tie. The most peculiar features are the appendages that adorn its head (Fig. 2.3), which give it a very characteristic appearance. These appendages are a reddish-brown, pyramid-shaped horny cap or crest and a pair of wattles, which are coral red in colour, at the base of the beak over a violet-black pigmented area.

The beak, which is 23–26 mm long, is horny, strong and curved, with a small caruncle above it that marks the boundary between the beak and the skull. The neck is smooth, with fine feathers spaced in the middle, forming a sort of sparse mane. The tarsi may appear either more or less pigmented in grey or blackish tones, and the hallux (rear-facing toe) is more elevated than the other toes. The grey-pearl plumage is uniform, bluish and shiny, with characteristic small white spots in the shape of lentils all over the body except for the chest and neck. The feathers have a serrated distribution on the wings and are larger toward the lower part of the plumage.

This bird can reach a live weight of over 2 kg as an adult, and because of its abundant plumage, which exceeds 100 g, it appears larger than it really is. Its skeleton is adapted for short, powerful flights, akin to the flights made by other similar birds such as the partridge or the pheasant, for which its bones, mainly the shoulder girdle, the vertebral column and the bones of the head among others, have evolved morphologically in order to reduce its weight. For their part, the bones of the legs, devoid of dewclaws, are very well adapted for walking and running.

Development

Females may lay more than 150 eggs (Fig. 2.4) per year, with lengths ranging from 44 to 58 mm. These eggs are smaller and more rounded than those of chickens, with a thicker shell and a yellowish or pale-brown colour with fine dark speckles, although they can sometimes be almost white. In comparison to chicken eggs, guinea fowl eggs are smaller but denser, with an average weight of 48–49 g. This characteristic is very interesting as it mitigates the problems of transporting them, requiring less space and presenting greater resistance to breakage and cracking.

The nest is usually located under a bush or within a clump of grass, in a shallow depression lined with straw and feathers. The typical clutch size ranges from 5 to 20 eggs, either in a single nest or in collective nests. The female is responsible for incubation, while the male defends the nest, both during the laying and incubation periods, a characteristic of monogamous birds. Incubation time ranges from 24 to 28 days, with an average of 27 days, at the end of which the guinea fowl hatch.

Newly hatched chicks (Fig. 2.5) are reddish-brown in colour, with five longitudinal black stripes on the head. The back is striped and spotted with black, the belly is yellowish and the legs and bill are red. These keets have filoplumes on the head and neck; these areas will become bare when the filoplumes are shed during the moult. At this stage, the skin covering these areas will become turgid and pigmented, beginning to acquire the colours typical of the adult bird.

About 7 weeks after birth, the guinea fowl reach full vigour and maturity. In adulthood, the female is heavier than the male and reaches sexual maturity earlier. This adult guinea fowl is highly resistant to inclement weather, except for extremely cold temperatures, which impair its development and stunt its growth, as has been

Fig. 2.4 Guinea fowl eggs, which are slightly smaller than chicken eggs, are easily identified by the fine dots on their thick, hard shells

Fig. 2.5 Newborn keets (*pintadones*) have five black stripes running lengthwise down their heads

observed on extensive farms in the province of Badajoz (Spain), where guinea fowl live outdoors. Their development is consistent throughout the year, except during periods when minimum temperatures reach around −3 °C and maximum temperatures reach around 44 °C.

Sexual Dimorphism

A very notable characteristic of these birds is the almost non-existent sexual dimorphism, making it very difficult to differentiate males from females (Fig. 2.6). With experience, the males can be identified by the presence of a wider, more developed horny crest and larger caruncles in males than in females.

Females have a horizontal back and loin, whereas in males, this region is curved, making them more upright and slender. Their size and weight are somewhat smaller than those of females, provided that both sexes belong to the same lineage and are bred under similar conditions.

In short, this sexual dimorphism is not very marked, being based on imprecise and often misleading characters (Ajenjo 1964; Hoffman and Völker 1969; Le Coz-Douin 1992).

Behaviour

The guinea fowl's original habitat is Africa, which is why it can be seen naturally in areas of this continent where the vegetation is sparse, but not totally absent. These birds can be seen moving in a dispersed manner, although forming groups, in arid plains with little brush where they mainly find their food or among the acacia

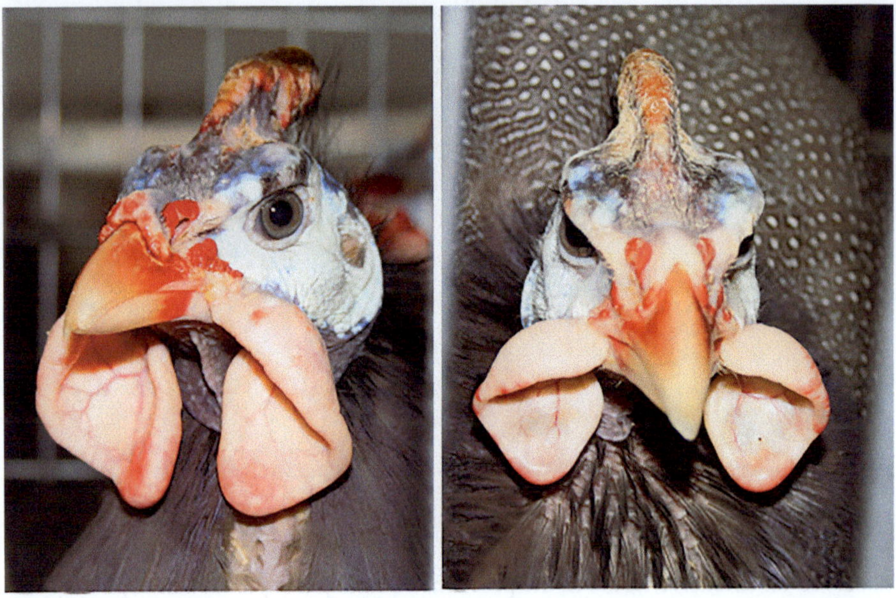

Fig. 2.6 Sexual dimorphism is not very noticeable, so it is not easy to differentiate between males and females

forests, which act as their main refuge. Guinea fowl feed on the ground but sleep in the treetops, protected from their predators. They can also be found in forests, either in wooded areas in valleys or on hills near watercourses, although this is only found in a small group of guinea fowl varieties. For the same reason, the most suitable climate for the guinea fowl is a dry, sunny climate with high temperatures, with a preference for dry, sandy soils.

Their diet is very varied; in the wild, they feed mainly on seeds and insects, although their diet varies according to the region and the season. In the spring and coinciding with the rain, they tend to feed on larvae and insects, while the rest of the year, they eat all kinds of seeds and small shoots of grass. Farm guinea fowl feed on poultry feed, supplementing their diet with other foods (seeds and insects) found in the farmyard when the farm is extensive, and they live in open-air enclosures.

The guinea fowl, both in the wild and domestic environments, has a highly gregarious nature, gathering in large groups that can range from twenty to several hundred individuals. They can only be seen forming separate pairs during the breeding season, as they are monogamous, as well as during the rearing period, when the pairs can be observed with their young of the year.

It should be noted that this is a bird with a very haughty and bold attitude and even, on many occasions, provocative, especially when it is in a confined situation. Nonetheless, it has a fearful reaction to any kind of danger, running away quickly or even escaping with short flights, especially when it is outdoor paddocks. In addition, they are highly sensitive to any strange being in their space whether in confinement or in pens, presenting a very curious reaction, as they immediately start to scream all together and loudly, with a very characteristic squawk. The males emit a sound similar to the syllables "cir, cir", while females emit a sound that could be represented as "coquah, coquah". For this reason, it is very advisable to locate the farm in a quiet area, away from roads and passageways, to maintain a peaceful environment that allows them to develop better (Belshaw 1985; Hart 1985; Le Coz-Douin 1992).

Competing Interests The author has no conflicts of interest to declare that are relevant to the content of this chapter.

References

Ajenjo C (1964) Enciclopedia de Avicultura, 2nd edn. Calpe, Madrid. 989pp
Belshaw RH (1985) Guinea-fowl of the world. Nimrod Book Services, Liss. 192pp
Clements JF, Rasmussen PC, Schulenberg TS, Iliff MJ, Fredericks TA, Gerbracht JA, Lepage D, Spencer A, Billerman SM, Sullivan BL, Smith M, Wood CL (2024) The eBird/Clements checklist of Birds of the World: v2024. Downloaded from https://www.birds.cornell.edu/clementschecklist/download/
GBIF Secretariat (2023) GBIF backbone taxonomy. Checklist dataset. https://doi.org/10.15468/39omei. Accessed via GBIF.org on 19 Aug 2024
Hart BL (1985) The behavior of domestic animals. W.H. Freeman and Company, New York. 390pp
Hoffman G, Völker H (1969) Anatomía y fisiología de las aves doméstícas. Ed. Acribia, Zaragoza. 190pp
Le Coz-Douin J (1992) L'élevage de la pintade. Editions du Point Vétérinaire, Maisons-Alfort. 252pp
Natural History Museum (2024) Natural History Museum (London) collection specimens. Occurrence dataset. https://doi.org/10.5519/qd.4gbppv6f. Accessed via GBIF.org on 19 Aug 2024. https://www.gbif.org/occurrence/1291469224

Predatory Characteristics of the Guinea Fowl (*Numida meleagris*) Against Pest-Forming Insects: Use of *N. meleagris* Against the Locust *Dociostaurus maroccanus*

3

José del Moral de la Vega, Ángel Mejías Guisado, Javier Jiménez Medina, Modesto Senero Fernández, and Jerónimo del Moral Martínez

Abstract

Locust infestations (*Dociostaurus maroccanus*) have been constant in the history of the Iberian Peninsula since prehistoric times because they are an essential part of some of its ecosystems. The most extensive of these plagues occur in Extremadura (Spain)—in the La Serena and Llanos regions—where, in 2011, the pest occupied an area of 400,000 ha. Industrially synthesised insecticides are necessary for its control. With the intention of reducing this dependence through biological methods, research projects were developed to test the effectiveness of birds as locust predators, whose carcasses could then be used by mankind for food. Previous studies to select the most suitable species indicated the use of *Numida meleagris* (guinea fowl). The research projects carried out proved the

J. del Moral de la Vega
Emeritus Head of the Department of Plant Pathology of CICYTEX, (Scientific and Technological Research Centre of Extremadura), CICYTEX-Finca La Orden, Guadajira, Badajoz, Spain

Á. M. Guisado · M. S. Fernández
CICYTEX (Scientific and Technological Research Centre of Extremadura), CICYTEX-Finca La Orden, Guadajira, Badajoz, Spain

J. J. Medina
FULGENCIO SPA, S.L., Granada, Spain
e-mail: fulgencio@fsagro.com

J. del Moral Martínez (✉)
Food Quality and Microbiology Research Group (AGA015) of the Extremadura Science, Technology and Innovation System, (Badajoz), Spain

J. del Moral de la Vega, J. del Moral Martínez (eds.), *The Guinea Fowl (Numida meleagris)*, https://doi.org/10.1007/978-3-032-11654-3_3

effectiveness of this bird as a predator and the feasibility of its management against the locust pest—adapted to the ecosystem, hardy, an efficient predator, easy to handle, and an ontogeny that aligns with that of the insect— as well as the excellence of the guinea fowl carcasses produced. This chapter describes the procedures developed in the field and in the laboratory.

Keywords

Guinea fowl · Locusts · Plague · Biological control · Semi-wild livestock farming

It can be generally stated that birds are great insectivores, and precisely because of this quality, by decimating insect populations, they prevent catastrophic infestations, one of the many reasons why man has always considered them beneficial.

Among the plagues that have most affected mankind are those of the locust, caused by an abnormal development of the population of various species of locusts; capable of causing famine in entire countries. To avoid these plagues or reduce their intensity, man has favoured the development of different animal species, including various species of birds. This procedure, although important, has never been decisive.

With the discovery of synthetic insecticides in the first half of the twentieth century, mankind managed to control many plant pests, albeit at the cost of releasing more or less polluting substances into the environment. Starting in the last third of the last century, medicine realised that the number of fatal diseases was increasing dramatically, and this seemed to be a consequence of the pollution of the environment by insecticides, which is why biological alternatives began to be sought. Among these alternatives was the use of birds whose carcasses could be consumed by humans—the insect is eaten by the bird, thus eliminating the pest, and man eats the bird, transforming a calamity into an added value of the rural world.

Within the most dangerous pests in Spanish agriculture are the locusts *Dociostaurus maroccanus*, an insect that is endemic in several Spanish regions, including La Serena (Badajoz, Extremadura), where it has been proven, for the first time in an exhaustive manner, that these pests can be effectively controlled by means of the appropriate management of the bird *Numida meleagris* (guinea fowl).

Locust Plagues in Spain Throughout Its History

In the earliest writings of the history of Spain, there are already references to the plagues of *Dociostaurus maroccanus* (Fig. 3.1). Saint Isidore does so in *The Etymologies*. In the ninth century, the Andalusian historian Ibn Hayyan wrote:

> In the year 845 and the following year there was a great drought that affected al-Andalus. Livestock were lost, vineyards were burnt, there was a plague of locusts, magnifying the famine and the difficulties of people's subsistence.

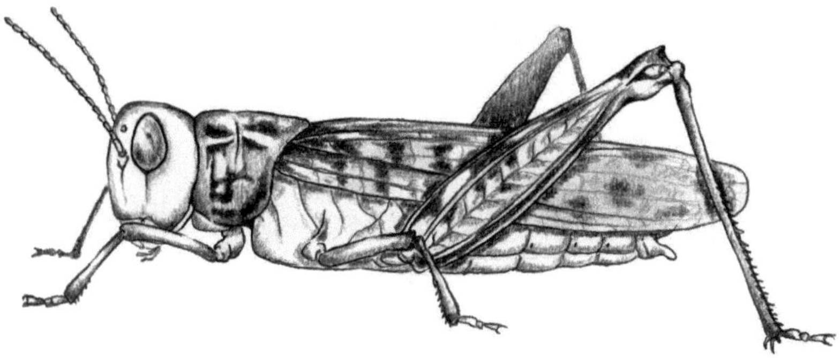

Fig. 3.1 Adult *Dociostaurus maroccanus*. (Original graphite by José Del Moral Martínez)

In the thirteenth century, Alfonso X the Wise established the use of the constellations as a warning to predict the imminent appearance of locusts (Navarro 1901; López Cordero and Aponte Marín 1993; Vázquez Lesmes and Santiago Alvarez 1993; Del Moral and Mejías 1999).

The disasters caused by this phenomenon were common to most Spanish regions, although their occurrence varied from year to year. In five of them (La Serena and Los Llanos in Extremadura, Los Pedroches in Andalusia, La Alcudia in Castilla-La Mancha, and Los Monegros in Aragon), these pests have been endemic and are still dangerous to this day, mainly in La Serena (Extremadura). In the 1930s, Moreno Márquez, one of the Spanish researchers who has most studied this phenomenon, wrote:

> The year I arrived in Badajoz, the locusts ruled the streets and houses of the city, making a shameful spectacle. (Del Moral and Mejías 1999)

In 1900, the official figures for the pest in Spain were 211,100 ha, and 110 years later (2010), the official area of action against the insect in Extremadura is 400,000 ha.

Agro-ecological Characteristics of La Serena, an Extremaduran Region with the Most Dangerous Locust Infestations

The district of La Serena is characterised by a uniform group of low hills with a broad base and rounded tops that occupy a large area, where pastureland predominates on sparse, poor soil and where the bedrock (composed of metamorphosed Precambrian materials) emerges in formations of what are known as "dogteeth" or "stone needles" (Fig. 3.2).

The climate in La Serena is dry and extreme, with winters that are not as harsh and summers that are hotter than the rest of the Iberian plateau. The rainfall regime is characterised by highly irregular rainfall. The average annual rainfall does not exceed 500 mm, with two rainy periods, one in spring (February–April) and another

Fig. 3.2 The soils of La Serena offer ideal conditions for the development of locusts. One of its most striking features is its "dogtooth" profile, a consequence of the geological folding of slate

in late autumn and early winter (October–December). The rest of the year is a long dry period from June to September, interrupted only by occasional thunderstorms.

Vegetation is almost exclusively represented by Poa and Festuca grasslands, which serve as food for livestock, mainly herds of merino sheep that produce wool, meat, and exquisite cheeses.

These characteristics of the La Serena region—shallow soils, a hilly profile, herbaceous vegetation cover, and low rainfall—have meant that agricultural production has been mainly oriented towards extensive sheep farming (Merino breed) and hunting (partridge), farming systems that justify the lack of cultivation to prevent erosion (Fig. 3.3). If we take into account that to complete its development, *Dociostaurus maroccanus* needs to be buried in the soil in clusters called "egg pods" (Fig. 3.4), from June–July to April (9–10 months), it is easy to understand why the characteristic agro-ecosystem of La Serena, so lacking in humidity, favours the endemism of this insect (Arias et al. 1993; Arias and Martínez de Velasco 1998).

Evolution of the Procedures Developed to Control Locust Infestations in Spain

Physical and Religious Procedures

The first documents we have on how to control locust plagues describe physical procedures, organising real armies armed with various devices, as we can read in this document from the sixteenth century, in which Doña Oliva Sabuco de Nantes describes "their strategy against the enemy":

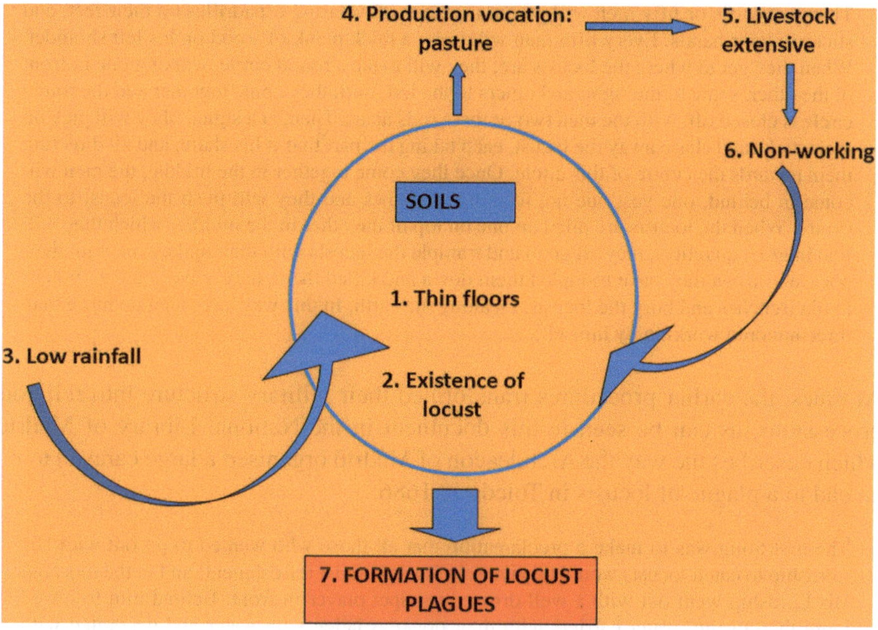

Fig. 3.3 Diagram showing the environmental and agricultural factors that favour the formation of locust pests in the La Serena region (Spain) and which are repeated in many other parts of the world

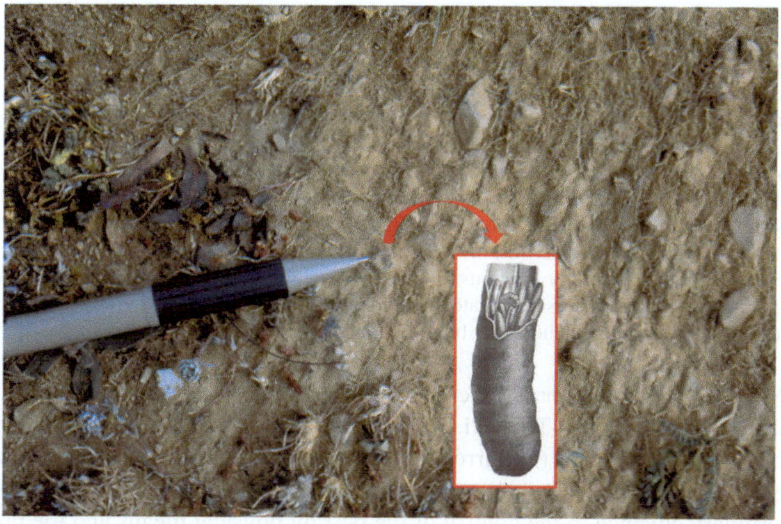

Fig. 3.4 Ootheca or "egg pods" buried in the soil in which the eggs of *D. maroccanus* are protected. The pencil marks the place where the operculum of an ootheca appears after a light layer of soil has been brushed away

> Thirty, or forty, or fifty men will go out together, all wearing espadrilles on their feet, and sticks in their hands. Every fifth man will carry a thick plank of wood on his left shoulder. When they get to where the locusts are, they will make a round circle, walking one in front of the other, some to the right, and others to the left, until they come together and the round circle is closed off, with the men two or three rods apart. Then, at a signal, they will pick up their sticks and chase away the locust, each taking the part that is his share, and all directing them towards the centre of this circle. Once they come together in the middle, the men will come in behind, one yes, one no, forming two rows and they will push the locust to the centre. When the locusts are piled up one on top of the other in the middle (which they will find in great quantity), they all go in and trample the locusts with their spikes and with their hoes, and when they have trampled them down and killed them, they leave some with hoes to dig trenches and bury the locusts, cleaning the earth. In this way, fifty men do more than three hundred working by himself....

At times, the earlier procedures transformed their military structure into religious processions, as can be seen in this document in the National Library of Madrid, which describes the way the Archdeacon of Madrid organised a large caravan to put an end to a plague of locusts in Toledo in 1686:

> The first thing was to make a proclamation that all those who wanted to go out with His Lordship to catch locusts would be given six Reales[1] and a refreshment; and in the morning His Lordship went out with a well-dressed trumpet player in front. Behind him followed more than twenty water carriers with their donkeys laden with water and decorated with poplar branches with many coloured ribbons. After these followed many wagons, and on each one a sign that said what carried, such as pantry, kitchen, pastry shop, confectionery, infirmary. Then followed a tent on a carriage drawn by four mules with tassels, and another tent where the people were to go to catch the locusts. Next came many mules loaded with feathers and weapons. Among the supplies there was one mule from which six silver horseshoes hung from the rack and on top of it a blacksmith's bag with its farrier's bench and horseshoeing equipment. Then followed a company of finely dressed soldiers firing many shots, and at the end of the soldiers one of them carried a cross seven quarters long with a very beautiful palm tree. After that came the crowd of the people, some on mules, others on donkeys. Then came the sworn knights and aldermen, all of them on very beautiful horses and were very gallant, and at the end of them was the aforementioned Archdeacon himself, with a horse decked out with black ribbons, wearing a short cassock, boots, and espadrilles (sic) and carrying a banner in his hand with the insignia of Saint Agustine, with his Doctor (sic), and surgeon and barber following behind. There were also six Augustinian friars on their mules, and many other religious figures who went to perform blessings. Behind these dignitaries was another horse decorated with many red ribbons and, on the saddle, a basket and a gilded hoe to dig the holes for burying the locusts....

In 1901, the efficacy of spraying concentrations of the insect with petrol and then burning them was proven. This led to the organisation of national campaigns organised by the State (Fig. 3.5) (Navarro 1901). On the other hand, the biological knowledge of the insect that was achieved at the end of the nineteenth century was an extraordinary advance in the control of the pest by physical means and the prophylactic effect of the winter ploughing carried out in the soil where the insect lays its eggs the previous summer (Fig. 3.6).

---

[1] Reales were coins used in Spain from the fourteenth century until the nineteenth century, when they were replaced by the peseta.

Fig. 3.5 Image from the 1940s showing a group of workers burning locusts with gasoline after bundling them with zinc trays and cattle chutes

Fig. 3.6 In the 1930s, ploughing the soil in Extremadura was an effective way of destroying locust populations in areas where locusts were endemic

Predation by Various Arthropod Species

Arthropod predators of locusts have been recorded in Spain since the end of the nineteenth century.

In 1886, the hymenopteran *Sphex paludosa* Rossi was observed preying on locust larvae, and in 1941, significant predation by the dipteran *Glossista infuscata* Meig and the coleopteran *Trichodes flavocinctus* Spin was recorded (Fig. 3.7). The

Fig. 3.7 Adult of the coleopteran *Trichodes flavocinctus* Spin, a common locust egg predator in La Serena, Extremadura (Spain)

action of the latter two species was accurately assessed in 1994, and each was found to be capable of destroying 20.4% and 3.4% of the oothecae under study.

Despite the predatory efficacy of these insects, in Extremadura, up to the present day, this phenomenon has developed in entirely naturally against locust plagues, without any standardised procedure having been developed for their massive and controlled use (Greathead 1992; Hernández-Crespo 1993).

Predation by Pigs and Birds

There are also historical references to the use of various species of animal predators of locusts (pigs, turkeys, chickens, etc.) for locust control (Toynbee 1973), as recorded in the minutes of the Ecclesiastical Chapter of Jaén in 1670. Even importing birds (*Paradisea tristis* L.) from the Philippines in the eighteenth century, where a great predatory efficiency of this species had been observed: "This bird, according to the information available, exceeds others in the hostility and tenacity with which it pursues the locust" (López Cordero and Aponte Marín 1993).

The insectivorous nature of birds has been valued throughout the history of these plagues in Spain, and the bibliographic archives are full of references to the effectiveness of numerous species (rooks, larks, thrushes, blackbirds, partridges, bee-eaters, orioles, sparrows, owls, vultures, kites). Leandro Navarro, at the beginning of the twentieth century, was astonished by the predatory capacity of storks:

> In the province of Cáceres, in the municipality of Arroyo del Puerco, and on a farm owned by Mr Germán Petit, we were struck by the extraordinary number of storks that nested in the various buildings on the aforementioned farm (Fig. 3.8). And there we saw how these very useful birds (one of the few respected in our country) devoured large quantities of locusts.

Fig. 3.8 House in Arroyo de la Luz—formerly known as Arroyo del Puerco, Cáceres (Spain), early twentieth century, with a roof full of stork nests

Although the bibliographical references on this biological control procedure are abundant, and all of them speak of its excellence, we have not found any that describe this procedure as a standardised technique for its development according to the variables of the pest and the agro-ecosystem where it appears (Heredia 1782; Navarro 1901; Del Cañizo 1939; Huddleston 1958; Greathead 1992; López Cordero and Aponte Marín 1993; Vázquez and Santiago 1993; Buj 1996).

Control Through Entomopathogens

Biological parasite control procedures by means of entomopathogens (viruses, bacteria, protozoa, nematodes, and fungi) are a very powerful means against numerous insect pests. They started to be used in Spain right at the beginning of the twentieth century (Waage 1990), although it is true that these procedures were nothing more than anecdotal references until the last decade of the twentieth century.

Among the viruses investigated against locust plagues in our country, some entomopoxviruses have been found, although with uncertain results. Certain strains of the bacterium *Bacillus thuringiensis* have been shown to be effective against the insect. As far as nematodes are concerned, the species *Hexamermis serenensis* is frequently found in the body of locusts in La Serena, although its importance has not been assessed.

In Extremadura, the most promising results have been obtained with fungi. Locusts with symptoms of "Summit disease" (Fig. 3.9), from which the fungus *Entomophaga grylli* (Fresenius) Batko has been isolated, are frequently found on several species of vines in La Serena.

In the region of La Serena, using biological traps to extract entomopathogenic fungi from the soil, some strains of *Beauveria bassiana* (Bals.) Vuill. have been isolated, with high virulence on *D. maroccanus* (Fig. 3.10).

Fig. 3.9 It is common to see in locust-infested areas some diseased or already dead insects on the top of thistles (Summit disease). This seems to be because locusts infected by the *Entomophaga grylli* fungus seek to store heat to limit or prevent the development of the pathogen, which is why they climb as high above the ground as they can

However, even though the use of entomopathogens to control insects has been proven to be effective and safe, they are limited by the fact that they need a precise temperature and humidity to act, and if this does not occur, the entomopathogen has no insecticidal value (Poinar and Thomas 1984; Greathead 1992; Hernández-Crespo 1993; Jiménez Medina 2000).

Control Through Industrially Obtained Insecticides

In Spain, chemical control has been and continues to be fundamental in the control of these *Dociostaurus maroccanus* pests since the beginning of the twentieth century. In 1901, the efficacy of spraying clusters of the insect with petrol and then burning them was proven, and this led to the organisation of national campaigns organised by the State (Fig. 3.11) (Navarro 1901). In those years, sprays of calcium arsenate (produced by dissolving sodium arsenate and lime in water) and food baits made by infusing bran with arsenite also began to be used. In 1948, Müller was awarded the Nobel Prize for discovering the insecticidal properties of DDT, thus initiating the industry of phytosanitary products known as chlorates, whose effectiveness, cost, ease of transport, and distribution were decisive in the control of insect pests. The insecticidal power of these products is very potent, and their price is very inexpensive, which is why until 1986, chlorinated HCH was used against locusts in Extremadura, being replaced by organophosphates from 1987, mainly fenitrothion and malathion.

At present, the pesticide used in Extremadura is diflubenzuron, an effective and highly specific insecticide, whose use, effectively controlled by the *Junta de Extremadura* (government of Extremadura), prevents the formation of dangerous locust infestations. However, this procedure neither prevents the release of large

Fig. 3.10 Nymph of *Dociostaurus maroccanus* Thunb covered with mycelium corresponding to the entomopathogenic fungus *Beauveria bassiana*

quantities of chemical products into the environment (Fig. 3.12), nor does it reduce the area where the parasite appears endemically. In fact, in 1901 (Navarro 1901), the extent of the pest in the province of Badajoz was 75,976 ha, whereas in 2010, intervention against the pest affected 400,000 ha.

Semi-wild Guinea Fowl (*Numida meleagris*) Breeding Project Aimed at Using Their Insectivorous Behaviour to Prevent Locust (*Dociostaurus maroccanus*) Infestations

There are abundant references to the use of predatory birds against locust pests, although this procedure has not been used as a routine phytotechnical measure where there are endemic plagues of the insect. To advance this technique in La Serena (Badajoz), two projects financed by the Regional Government of Extremadura, INIA (Ministry of Science and Innovation) and ERDF (EU), were

Fig. 3.11 In the first half of the twentieth century, locust infestations were controlled by means of campaigns in which engineers and workers were organised as if they were military operations. Image showing a team of technicians in their bivouac before starting a day's work in the spring of the 1940s

Fig. 3.12 The application of synthetic insecticides leads to environmental pollution as a secondary problem. Remains of insecticide containers used in a campaign against locusts in Extremadura (Spain) in 1992

developed. After screening a number of insect predatory birds whose carcasses are used in human food, four species were chosen: the turkey (*Meleagris gallopavo* L.), the partridge (*Alectoris rufa* L.), the hen (*Gallus gallus* L.), and the guinea fowl (*Numida meleagris* L.) (Fig. 3.13). The research was carried out on the La Gama

Fig. 3.13 Guinea fowl chick and adult (*Numida meleagris* L.)

farm (500 ha) located in the region of La Serena (Badajoz), a farm that is affected by locust infestations every year. After 5 years of study, it was possible to design a semi-wild breeding system for the "painted bird" for locust predation, a technique that is economically viable and which produces carcasses usable for human consumption.

Demonstration of the Good Adaptability of Guinea Fowl to the Agro-ecosystem of La Serena

The studies carried out to determine the possibility of using the guinea fowl against locust pests were based on the hypothesis that this bird, being of African origin, should be well adapted to the climatic conditions found in La Serena. However, it is possible that the batches of birds used at the beginning of the studies had already lost these hardy qualities. For this reason, before starting the experiments on locust predation, other experiments were designed to test the ability of the guinea fowl to adapt to the maximum and minimum temperatures that usually occur in the south of Extremadura. To this end, during 1 year, in each of the four climatic seasons, the behaviour of 100 birds reared in total extensively (Fig. 3.14) was compared with the same number of birds protected indoors. The extensively reared guinea fowl had no protection other than a plastic mesh shade structure for the summer and an inverted U-shaped metal sheet on the floor for winter roosting. The birds in each flock were slaughtered at 16 weeks of age and the behaviour of the animals and the weight of the carcasses were determined.

From a livestock farming perspective, it was found that the carcass yields of the birds reared outdoors decreased significantly in the periods of maximum and minimum temperature and humidity (43.5 °C temperature and 44% relative humidity;

Fig. 3.14 Guinea fowl reared extensively in Extremadura

−3 °C and 77% relative humidity). That being said, all the birds withstood these conditions, showing that the adaptation of the guinea fowl to the agro-ecosystem existing in La Serena is similar to that of any of the birds existing in the south of Extremadura.

Characteristics of Predatory Guinea Fowl and the Facilities Needed to House Them

These birds, when they develop their true feathers, have a great capacity for flight, and it has been found that a dog used for hunting birds is incapable of capturing them without the help of the hunter. To prevent them from flying high and escaping from their confinement enclosures, it is advisable for the chicks or keets to have their wings clipped.

Guinea fowl intended for locust predation are brought to the field to feed on the insects when they have fledged and are well adapted to the ambient temperatures—about 6 weeks of age.

For the breeding of these locust-predatory birds to work effectively, it is necessary to coordinate the appearance and evolution of the insect with the development of the birds. This ensures that when the plague is about to form, the guinea fowl are perfectly adapted to the conditions in which they will live and can fatten by feeding on the insect. When the locust disappears at the end of its life cycle, the birds will have reached a suitable weight (approximately 1.5 kg) to be removed and taken to the slaughterhouse for processing and subsequent distribution on the market. In La Serena, the region where these studies were carried out, the time that the birds

Fig. 3.15 It is recommended that the construction of shelters for predatory birds should consist of lightweight modules (wire mesh) that are easy to transport and assemble

remain in the field feeding on locusts was from the beginning of April to the end of June.

The infrastructure needed to house and care for the birds in the field should be defined by the following characteristics: easy to transport, without the need for additional installation work, easy to dismantle and store, and made of low-cost durable materials designed to last 4–5 years (Fig. 3.15).

The pens or shelters where the birds are to be housed can be made up of square modules of 2 × 2 m made of a frame of light metal sheeting and poultry netting. These segments are attached to each other by means of wire or rope to form a square, so that the established structure, without the need for any anchoring or foundations, remains firm, serving to enclose the birds during the night and defend them from possible predatory mammals (foxes, genets, ferrets, etc.). A shelter of 100 m^2 can serve for 500 birds.

To protect the birds from the cold at night or from the rain, it is recommended to have some metal roosts inside the shelter, in the shape of an inverted U, 2 m long, 1 m wide, and 0.8 m high, open at the ends and resting directly on the ground. It is also advisable to cover the ground with a bed of straw to avoid humidity and to collect the birds' droppings, changing the straw every 3–4 weeks. Five roosts of these characteristics are sufficient to house the birds in the 100 m^2 refuge.

In order to protect the guinea fowl from excessive afternoon heat during the month of June, 2-m-high shady structures with a 3 × 3 m plastic mesh covers should be erected. Eight or nine units in each shelter are sufficient to shade the birds.

Inside the shelter, hopper-type feeders, made of zinc plate and with a capacity of 12 kg each, must be provided.

For drinking water for guinea fowl, a 2-m-long drinking trough with a constant level per buoy should be provided, connected to a water tank with sufficient capacity to ensure that the birds can be supplied for 4–5 days without refilling (Fig. 3.16).

Fig. 3.16 Corral built in La Serena, Extremadura (Spain), to protect predatory guinea fowl at night

For the birds' comfort, a wooden perch with an inclined plane can be placed inside the shelter, although this is not essential (Castelló 1996; Buxade 1998).

Bird Handling

When the guinea fowl arrive in the field, they must adapt to their new environment (temperature, humidity, light) and learn to locate their shelter or pen so that when they go out into the field, they can freely prey on the insect. To do this, for 1 or 2 weeks, they are kept permanently enclosed in their shelter, eating and drinking ad libitum. After this time, the guinea fowl will have "memorised" the location of the pen, being able to move away from it for up to 1 km and return on their own.

The pen opens at dawn and remains open until dusk. The guinea fowl leave the pen in groups of a hundred birds or more as soon as the gates are opened (Fig. 3.17). Once outside, they begin to explore the ground in order to locate areas where the insect is present, an operation in which they can be guided by the keeper. It is very common for the flocks of guinea fowl to move across the field, forming a large line that moves forward detecting locust hotspots. When one of them locates a cluster of the insect, the birds regroup, capturing them with great efficiency. When their crops are full, which usually takes about 2 hours, they return to the refuge to drink water and rest. They remain in or around the refuge until the afternoon (4–5 pm), then they go out again to capture locusts, returning to the pen when the sun begins to set, at which time the keeper closes the doors of the refuge to protect them from birds of prey. The interest of this semi-wild guinea fowl farming lies in them fattening up on the locusts they catch. However, the density of locusts varies greatly from 1 day to

Fig. 3.17 When the pen doors open at dawn, the guinea fowl quickly leave the pen. They come out in groups and, either on their own or guided by a keeper, locate the insects and capture them with great efficiency

another during the 3 months that there are insects in the field, so it is common for the birds have a lot of prey to hunt on some days, while on others they have hardly any. On the other hand, birds have a remarkable ability to locate insect hotspots, but this ability is not absolute; there are days when, during their foraging, they only catch a few insects, such as ants or occasionally a cricket. To keep their predatory instinct sharp while meeting their nutritional needs, the birds are given a supplementary ration of feed but only once they have returned to the pen at dusk and the keeper has secured them inside. This way, the birds consume more or less feed depending on how much they have satisfied their hunger with locusts—their preferred food. In the morning, when the pen doors are opened and the birds stampede out, the keeper hangs up the feed hoppers, so that no birds have any feed until the evening, and those that might be left behind in the pen have no feed at all during the day, thus inviting them to go out and hunt.

In investigations carried out to compare the feed intake of guinea fowl that were totally confined (Fig. 3.18) with that of those that were free and caught locusts, it was observed that, while the former ate more than 150 g of feed/bird per day, the latter barely exceeded 50 g of feed/bird. The final result of the experiment, after 48 days, was that those that were constantly confined and ate feed ad libitum weighed 1884 g/bird, while the free-range birds weighed 1744 g/bird, although this difference was not statistically significant. The carcasses produced were analysed using a method developed by the Working Group of WPSA for Sensory Analysis of Broilers (INRA 1975, 1985, 1989; Blum 1979). Six professional tasters determined, by blind tasting, the flavour, juiciness, tenderness, fatness, and overall quality. They determined that although the overall value of all carcasses was similar, the flavour of meat from guinea fowl that had freely eaten locust was superior to meat from birds that had been fed only fattening feed and had been kept indoors throughout their lives.

Fig. 3.18 Pens designed to conduct one of the guinea fowl feeding experiments

Fig. 3.19 Locust band with a high density of locust nymphs

Guinea Fowls' Ability to Explore Beyond Their Shelter or Corral

One of the characteristics that should be required of a predatory bird is that it should be brave enough to stray far from its shelter and freely locate, without the need of a keeper, the insect's swarms or bands (Fig. 3.19).

Fig. 3.20 Although guinea fowl have a great ability to locate locust lines, if a keeper points them in the right direction, locating time is dramatically shortened

The guinea fowl show a great capacity for exploration, moving up to 1 km away from their refuge. They can do this without the need for a caretaker, although a little guidance can help direct the birds to one part of the farm or another (Fig. 3.20).

In the studies carried out in Extremadura, to determine as objectively as possible the areas affected by locust infestation and their predation by the birds, the caretaker recorded the main daily routes taken by the birds from the farmyard throughout the duration of the locust plague. These routes were stored on a computer, and at the end of the campaign, all were superimposed, as shown in Fig. 3.21. The area of pest control during the two-and-a-half month infestation was estimated at 314 ha.

The Guinea Fowl: A Good Locust Predator

In order to test the predatory efficiency of the guinea fowl, researchers from the *Junta de Extremadura* have carried out an experiment to evaluate this capability compared to that of chickens, a bird whose insectivorous characteristics are well known in the region of La Serena.

The experiments were carried out on a farm in La Serena, where there was a pest density of 4–5 locusts/m^2 and some outbreaks of locust bands. In the centre of the farm, there were two refuges with 500 birds each, one with hens (*G. gallus*) and the other with guinea fowl (*N. meleagris*).

When morning temperatures were high enough for the locusts to exhibit normal activity, three batches of ten birds of each species were taken and transported in cages to three different areas where the insect was clearly present (three blocks of ten birds of each species). The birds, once released from their transport cages, ate freely for 2 hours. After this time, they were collected and slaughtered. The contents of their crops were analysed with the aid of a stereomicroscope and weighed using a scale to the nearest 0.1 g. These tests were carried out five times during the spring.

Statistical analysis (ANOVA) of the results obtained showed that, in the 2 hours of the experiment, guinea fowl were able to ingest a much greater weight of food

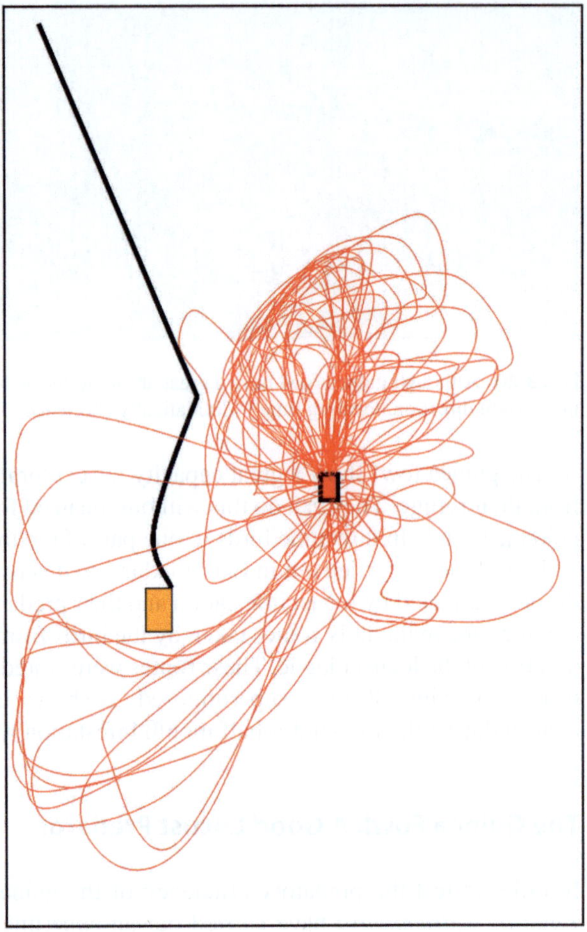

Fig. 3.21 Drawing showing the daily routes followed by the guinea fowl in their predatory function around their shelter during the months of the plague. On the left, farmhouse and farm road; in the centre, a pen

than chickens (Fig. 3.22), mainly locusts—the relative value of the weight of locust ingested by guinea fowl in the total amount of food varies between 89% and 99% (Fig. 3.23). The differences between the two values are highly significant (99% confidence). These results coincide with observations made in the field, where guinea fowl are more agile than chickens in capturing insects.

In order to know precisely what type of food the guinea fowl ate during the time the locusts were in the field, five guinea fowl were randomly taken weekly from among 500 birds in the pen and slaughtered after predatory activity (2 hours). Food intake was examined using a stereomicroscope, determining for each crop the weight and number of locusts ingested, as well as coleopterans, dipterans, hymenopterans, lepidopterans, seeds, plant remains, and other elements found. The number of locusts parasitised by the mermithid nematode *Hexamermis serenensis* (Hernandez-Crespo & Santiago-Alvarez, 1997), and the existence of adults of *Glossista infuscata* Meig. and *Trichodes flavocinctus* (Spinola, 1844) were specifically looked for, due to their importance in the control of locusts. Figure 3.24 shows

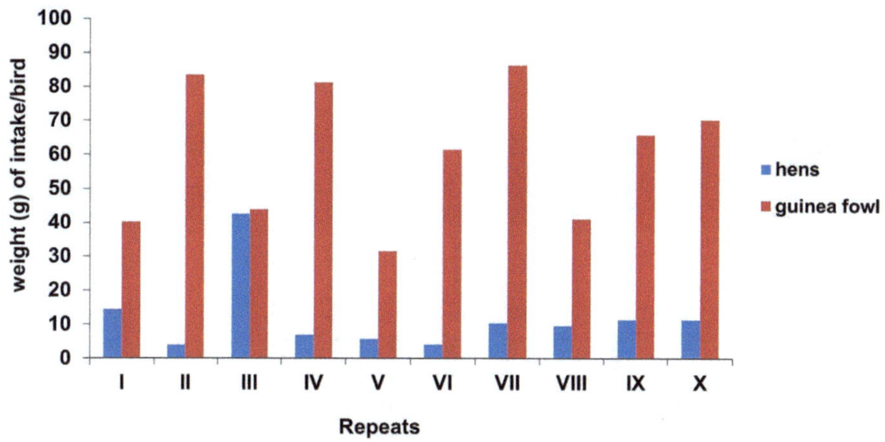

Fig. 3.22 Mean weight of locusts ingested by birds in an experiment designed to test the predatory efficiency of guinea fowl compared to chicken hens

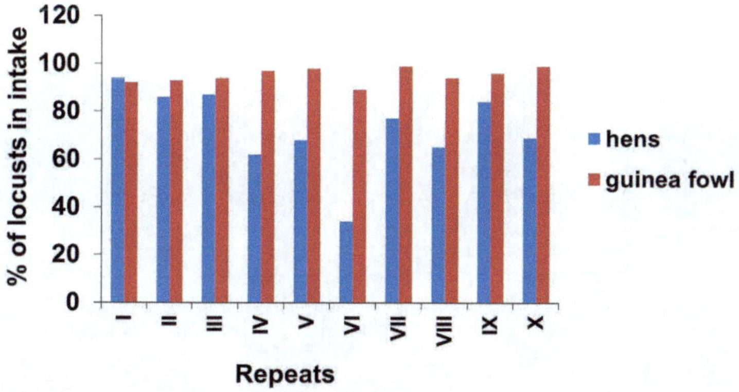

Fig. 3.23 The preference of guinea fowl (red) for eating locust is very high; the relative value of the weight of locust found in the birds' crop to the total food ranges between 89% and 99%

the relative value (%) of insects and plants with respect to the total intake of the guinea fowl over 8 weeks of observation. It shows a clear preference of the bird for insects, and within the plant material, it is seeds that it chooses, from which it can be inferred that this bird behaves not only as an insectivore but also as a granivore.

The relative value of locusts predated during the months of May and June with respect to insects of other orders varies greatly (88.05% were locusts, 6.32% were coleopterans, 4.72% lepidopterans, 0.62% dipterans, and 0.28% hymenopterans). Locusts are the insect most frequently caught by the guinea fowl, and although captures of other orders are also observed, their value is much lower than that of locusts. Within the Coleoptera preyed upon, the *Trichodes* spp. stand out—accounting for 9% of the total number of Coleoptera. Among the locusts in the birds' crop, 14 of them—0.4% of all locusts eaten—were parasitised by the nematode *Hexamermis* spp. (Fig. 3.25).

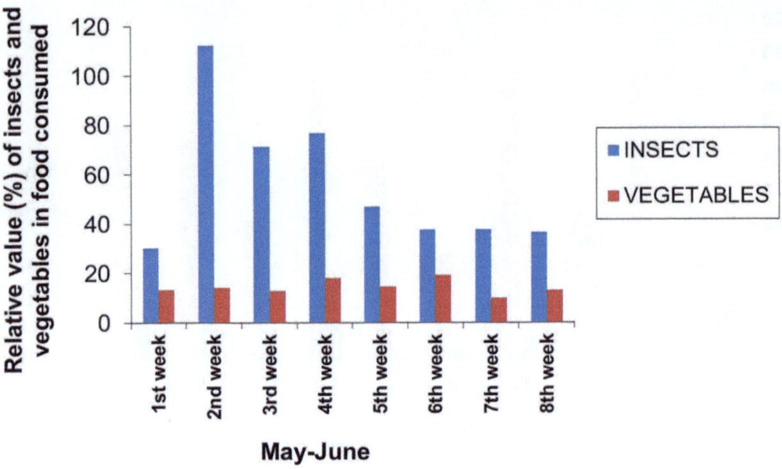

Fig. 3.24 Bar chart showing the preference of guinea fowl for insects as a food source

Fig. 3.25 Locusts found in the crop of a guinea fowl where a nematode (*Hexamermis serenensis* Hernandez-Crespo & Santiago-Alvarez, 1997), parasitising it, can be seen coming out of the broken abdomen of one of the locusts

From the studies carried out, it can be affirmed that this bird is mainly insectivorous and, in particular, has a great locust-catching capacity, existing some guinea fowl that, in 2 hours of free predation, are capable of ingesting 299 locusts.

Coordination Between Insect and Predatory Bird Development

In the semi-wild guinea fowl farming project to control locust infestations, the evolution of the insect and the bird are synchronised, so that when the insect hatches, the guinea fowl are ready to prey freely. The birds, 6 weeks old and already feathered, are taken to the field at the beginning of April when the presence of locusts begins to become widespread and remain there until the insect disappears from the field. The guinea fowl grow as they feed on the locusts, and when the surviving insects lay their eggs and disappear (generally at the end of June), the birds have the weight required by the market to be slaughtered (carcass of 1.5–1.7 kg) (Fig. 3.26).

The number of locusts preyed upon by the birds in 2 hours of observation varies throughout the test period as the intensity of the pest in the field changes, with a decrease in the number of locusts captured coinciding with the increase in size of the insect as it becomes an adult (Fig. 3.27).

Assessment of the Usefulness of Semi-wild Guinea Fowl Farming Against Locust Plagues

In order to design this project for the control of locust infestations by means of a semi-wild guinea fowl flock, numerous experiments were carried out, tests that

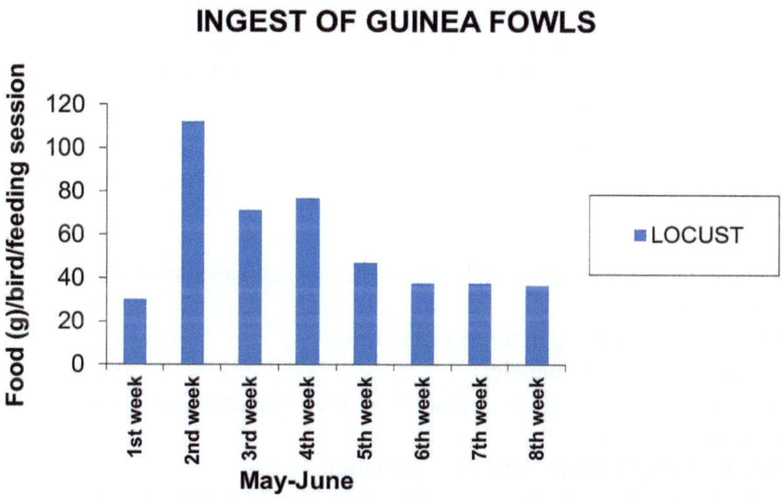

Fig. 3.26 The number of locusts caught by guinea fowl varies over the predation time, varying with the intensity of the plague

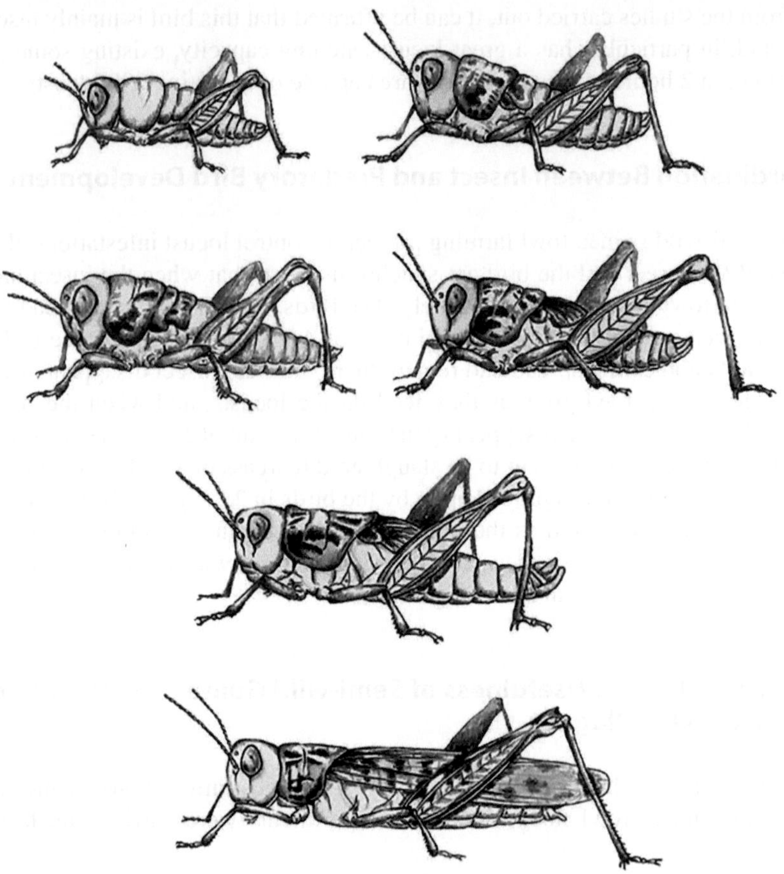

Fig. 3.27 Morphological characteristics of each of the developmental stages of the locust (*Dociostaurus maroccanus*). (Original graphite by José Del Moral Martínez)

were suitably designed so that the results obtained could be analysed mathematically. These results allow the following conclusions to be drawn about the functionality of the guinea fowl:

- They can be considered well adapted to the agro-ecosystem where locust infestations develop.
- They are undemanding in terms of facilities.
- They are very easy to handle.
- They have a good ability to explore from their shelter or pen (>300 ha) without the need of a keeper.
- They have a high ability for locust predation.
- Where locust infestations occur, they show considerable selectivity to prey on locusts, as opposed to other food sources.

- It is easy to coordinate their growth rate with the evolution of the locust infestation.
- The meat of guinea fowl reared in this semi-wild locust-predator farm has outstanding organoleptic characteristics.

However, any experiment, as a result of the equipment required to carry it out, has a disturbing effect on the results, which can lead to false hypotheses being considered valid. In order to avoid or mitigate this phenomenon, the experimental findings obtained have been applied to large farms (500 ha) in the region where these pests occur, verifying that this procedure for biological control of locusts is effective, feasible, and economically profitable. It can therefore be used as an alternative or complementary approach to pest control using insecticides.

The use of this semi-wild guinea fowl farming against locusts is an unprecedented method, although it is essentially an adaptation of the predatory phenomenon routinely observed in storks and other birds in the Extremaduran region of La Serena (Fig. 3.28), a natural and beautiful phenomenon which, if it is to be preserved, forces us to gradually replace synthetic insecticides with biological methods such as this, procedures that not only fulfil their purpose of controlling locust pests but also add considerable production of guinea fowl carcasses, a meat that is highly valued by European consumers and of great economic value (Del Moral et al. 1997, 1998a, b, c, 1999, 2004; Mejías et al. 1999; Tejerina et al. 2009).

Competing Interests The authors have no conflicts of interest to declare that are relevant to the content of this chapter.

Fig. 3.28 Storks and Great Bustards preying on locusts at the foot of Magacela, Badajoz, a common biological phenomenon in Extremadura (Spain) that has served as a reference for the design of a semi-wild farming system to combat locust infestations

References

Arias A, Martínez de Velasco D (1998) The locust in Extremadura. Agricultura 796:914–918

Arias A, Alvez C, García F, Martínez de Velasco D, Olivera J, Prieto A, Santos R (1993) Control of the Moroccan locust (*Dociostaurus maroccamus* Thunb) in Extremadura during the decade 1983–1992. Bol Sanid Veg Plagas 19:425–453

Blum JC (1979) Les besoins alimentaires et la composition corporelle du pintadeau. L'élevage de la pintade. Session ITAVI, May 1979

Buj A (1996) El Estado y el control de plagas agrícolas. MAPA, Madrid, 348 pp

Buxade C (1998) Zootechnics. Bases of animal production. Monograph II. Alojamientos e instalaciones II. Editorial Mundi-Prensa, Madrid

Castelló JA (1996) Exploitation of guinea fowl. In: Buxadé C (ed) Zootecnia. Bases de producción animal, vol X

Del Cañizo J (1939) Plagues of locusts in Spain. Bol Pat Veg Ent Agric 8:27–48

Del Moral J, Mejías A (1999) Life and work of Víctor Moreno Márquez. Extremaduran researcher in entomology and agronomy. Provincial Council of Badajoz

Del Moral J, Mejías A, Cancho M, Bosch A, Muriel A, Solana A, Jiménez J, López F (1997) Locust (*Dociostaurus maroccanus* Thunb.) pests in Extremadura. First results of their control by hen chickens. Phytoma-Spain 86:45–50

Del Moral J, García-Cuadrado N, Mejías A, Corrales D, Cancho M, Jiménez J (1998a) The Extremadura blue hen and the guinea fowl: predators of locusts. Agric Rev Agropecu 796:920–922

Del Moral J, Mejías A, Jiménez J, Cancho M, García N (1998b) From the locust plague to the Extremadura blue hen and the guinea fowl. Selecciones Avícolas XL(5):263–272

Del Moral J, Mejías A, Jiménez J, Cancho M, García N, Corrales D (1998c) The Extremaduran blue hen and the guinea fowl. First results of experiments carried out with these two birds to test their predatory power on locusts in south-west Spain. Agriculture LXVII(796):920–922

Del Moral J, Mejías A, Jiménez J (1999) Efficiency of the guinea fowl (Numida meleagris L.) as a locust predator (*Dociostaurus maroccanus* Thunb.). In: Congreso Nacional de Entomología Aplicada. VII scientific conference. Ed. Junta de Andalucía, 109 pp

Del Moral J, Miguel-Lasobras EM, Álvarez IS, Muriel A, Pérez-Rojas F, Senero M (2004) Toxicity in birds (Numida meleagris L.) of insecticides used in locust (*Dociostaurus maroccanus* Thunb.) control. Bol San Veg Pests 30:635–648

Greathead DJ (1992) Natural enemies of typical locust and grasshoppers: their impact and potential as biological control agents. In: Prior C, Lomer CJ (eds) Biological control of locust and grasshoppers. CAB International, Wallingford, pp 105–121

Heredia V (1782) Sobre remedios para libro de las langostas. In: Memoria del Consejo Provincial de Fomento de Zaragoza, 1914–1924, p 10

Hernández-Crespo P (1993) The Mediterranean locust, *Dociostaurus maroccanus* (Thunberg, 1815), its native natural enemies and the possible control of its pests by pathogenic microorganisms. Doctoral thesis. School of Agricultural and Forestry Engineers, Córdoba

Hernandez-Crespo P, Santiago-Alvarez C (1997) Description of *Hexamermis serenensis* sp. n. (Nematoda: Mermithidae), parasite of *Dociostaurus maroccanus* (Thunberg) (Orthoptera: Acrididae) in Spain. Fundam Appl Nematol 20(1):37–42

Huddleston JA (1958) Some notes on the effects of bird predator on hopper bands of the desert locust (*Schistocerca gregaria* Forskal). Entomatologist's Mon Mag 94:210–214

INRA (1975) Institut national de la recherche agronomique. Les volailles de consommation. Séance de travail 16 et 17 Octobre 1975. INRA éd. Station de recherches avicoles, Monnaie

INRA (1985) Institut national de la recherche agronomique. Feeding of guinea fowl. INRA éd. Station de recherches avicoles, Monnaie

INRA (1989) Institut national de la recherche agronomique. L'alimentation des volailles. Le pintadeau de chair. Juin 1989. INRA éd. Station de recherches avicoles, Monnaie

Jiménez Medina J (2000) Evaluation of isolated acridid fungi in the "reserve area" of La Serena for the control of pests of *Dociostaurus maroccanus* (Thunberg, 1815). Doctoral thesis. School of Agricultural and Forestry Engineers, Córdoba

López Cordero JA, Aponte Marín A (1993) A terror over Jaén. Las plagas de langosta XVI–XX. Ed Ayuntamiento de Jaén, 294 pp

Mejías A, Del Moral J, Jiménez J (1999) Discussion on the practical and economic feasibility of using guinea fowl (*Numida meleagris* L.) to control locust pests. In: National congress of applied entomology. VII scientific conference. Ed. Junta de Andalucía, 110 pp

Navarro J (1901) Natural means of extinction of the locust plague. Memoria dela Campaña contra la langosta. 1900–1901. Ministry of Agriculture, Industry and Trade, Madrid, 178 pp

Poinar GO, Thomas GM (1984) Laboratory guide to insect pathogens and parasites. Plenum Press, New York

Tejerina D, Lopez-Parra MM, García-Torres S (2009) Potential use of near infrared reflectance spectroscopy to predict meat physico-chemical composition of guinea fowl (*Numida meleagris*) reared under different production systems. Food Chem 113(4):1290–1296

Toynbee JMC (1973) Animals in Roman life and art. Cornell University Press, Ithaca; New York

Vázquez Lesmes R, Santiago Alvarez C (1993) Locust plagues in Cordoba. Ed. Cajasur

Vázquez R, Santiago C (1993) Locust plagues in Cordoba. Ed. Cajasur, 231 pp

Waage J (1990) Ecological theory and selection of biological control agents. In: Mackauer M, Ehler LE, Roland J (eds) Critical issue in biological control. Intercept, Andover

Reproduction and Breeding 4

Pedro Emilio Rosado Salgado, Francisco Pérez Rojas, and Modesto Senero Fernández

Abstract

This chapter deals specifically with the breeding of guinea fowl, with the aim of simplifying and making the process more efficient. In the interest of producing guinea fowl all year round, a breeding procedure by artificial insemination has been chosen, allowing, in addition to being able to select the best breeders and good yields in terms of chick production and to ensure the sanitary conditions of the whole process. The method chosen to extract the semen from the cockerels was the dorsal-abdominal massage method developed by Burrows and Quinn, with particularities derived from the experience gained in this task over years of experience. With regard to the eggs obtained, a detailed health programme for hatching eggs is specified, as well as the conditions and management of the incubator in order to obtain the chicks. The minimum ratio of non-viable chicks to healthy chicks is given in a table specifying the corresponding aetiologies.

Keywords

Artificial insemination · Eggs · Hatching · Breeding · Culling

P. E. Rosado Salgado (✉)
Locust Biological Control Working Group, Department of Plant Pathology, CICYTEX (Scientific and Technological Research Centre of Extremadura), Guadajira, Badajoz, Spain
e-mail: perosados01@educarex.es

F. P. Rojas · M. S. Fernández
CICYTEX (Scientific and Technological Research Centre of Extremadura), Guadajira, Badajoz, Spain

J. del Moral de la Vega, J. del Moral Martínez (eds.), *The Guinea Fowl (Numida meleagris)*, https://doi.org/10.1007/978-3-032-11654-3_4

Breeding Guinea Fowls

Battery Cages and Lighting Programme for Breeders

Birds used for breeding are housed in the battery cages at 23 weeks of age (Fig. 4.1) with a light programme of 14 h and an intensity of at least 15 lux. The females reach a laying rate of 50% at about 34–35 weeks of age (Fig. 4.2), with maximum laying at about 40 weeks of age and a gradual decrease at 70 weeks of age, at which time the breeding flock is removed and replaced by a new flock.

In the case of the males, the process is similar. They have the same light and housing conditions as the females. They reach sexual maturity at 30 weeks of age, and then they are removed from the batteries at the same age as the females.

Male guinea fowl will have their own selection system, eliminating all those with a weight 15% above or below the considered standard, choosing those with strong and straight legs and toes, bright eyes, bright colours on their wattles and the skin of the head, straight keel and an absence of spine defects, as well as good development of the penis (free of defects).

As complementary but unavoidable aspects of brood stock management, the following should be considered:

- A lot of cleaning is necessary in the breeder room to avoid accumulation of ammonia gases due to lack of litter renewal.
- The temperature must be controlled so that it is appropriate at all times, depending on the time of day and year.
- The presentation of the feed should be as homogeneous as possible in order to achieve uniform and effective consumption.
- Sanitary control of breeding stock to prevent pathological processes (Leclercq 1982; Le Coz 1992; Sauveur and Reviers 1992; Castelló 1996).

Fig. 4.1 Breeders housed in their cages

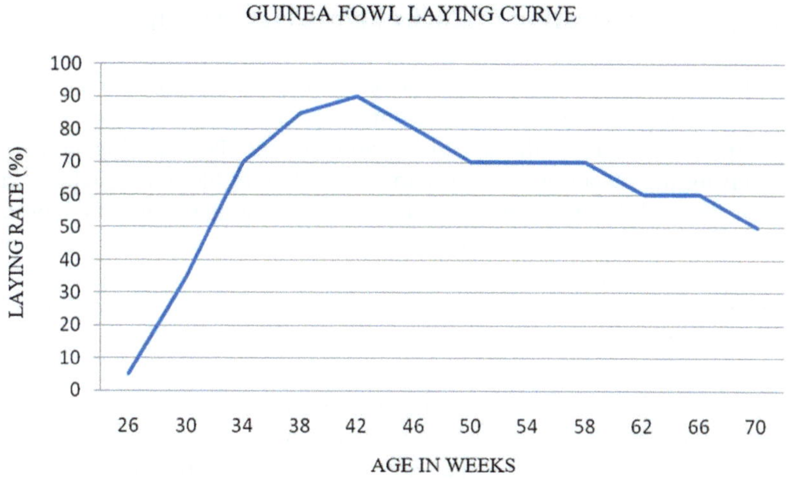

Fig. 4.2 Curve of guinea fowl deposition over its lifetime

Artificial Insemination

The technique of artificial insemination in the guinea fowl consists, broadly speaking, of extracting semen from the males and using it to immediately fertilise the breeding females. It is used in birds such as the guinea fowl, where sexual behaviour tends towards monogamy. This technique also makes it possible to increase the reproductive performance of the males, reducing the number of males per female, since, by dividing the semen into a certain number of insemination doses, a greater number of guinea fowl can be inseminated with the semen of the same male. In addition, this method of reproduction makes it possible to produce guinea fowl all year round, whereas under natural conditions, guinea fowl undergo a sexual rest dependent on variations in the photoperiod.

Insemination results depend on three important factors:

1. Quality of the breeders
2. Technique used
3. Farm health

Artificial insemination has been used in poultry for many years, in research work, in breeding programmes and in the production of poultry for the meat trade. Another advantage of artificial insemination is that it improves farm health and production control.

The technique by which semen is extracted from the cockerels is called the dorsal-abdominal massage technique or Burrows and Quinn technique (1937, 1939). The semen is then deposited, fresh or diluted, into the vaginal area of the female to obtain fertile eggs.

Training and Preparation of Cocks Before Harvesting

This consists of familiarising the male and female guinea fowl with the manipulations carried out during the artificial insemination technique. This is a very important period which allows the synchronisation of all the breeding animals and homogenising the response of the group of breeders to the manipulations carried out. In cockerels, it has the added advantage of stimulating sperm production.

The training of roosters used as breeders is carried out over the first few weeks following their introduction into the battery cages and consists of performing the same manipulations that will be carried out during semen collection (Burrows and Quinn 1937, 1939). Generally, after 3–4 weeks of training, the cockerels are ready to start producing adequate quantities of semen for insemination. In the case of females, it is advisable to do the training activities as well, as this reduces the stress level of the females when they are being inseminated.

During this period, animals that are not suitable as breeding stock should be culled.

Evaluation of Breeding Guinea Fowls

In order for artificial insemination to be successful, both the males and the breeding females have to be evaluated. In this evaluation, the general condition, the conformation, the state of the genital apparatus in both sexes, the response to stimulation by dorsal-abdominal massage in the case of the males (libido), the correct production of semen and its quality are considered.

To evaluate the quality of semen produced by cockerels, the macroscopic characteristics must be taken into account, volume and colour of the semen produced, as well as the microscopic characteristics, morphology, sperm motility and concentration of the spermatozoa (Lake 1971).

Semen Extraction Technique in Guinea Fowl

The technique of semen extraction in guinea fowl is based on the technique developed by Burrows and Quinn (1937, 1939). This manual technique is based on the ejaculation triggering action of the posterior dorsal-abdominal massage, by stimulating the erogenous nerve endings found in the cloaca area and in the line connecting the ischia, continuing along the keel line in the sternum.

Semen collection normally requires two people, but if a collection bank is used, only one person is needed. In order to speed up the operations, reduce stress on the birds and obtain cleaner semen, it is recommended that two trained operators are involved.

One of the handlers shall remove the guinea fowl from the battery cage; this operation shall be carried out causing as little stress to the animal as possible. Painted cockerels can inflict painful pecks, so it is important that the handler in

Fig. 4.3 When removing males from their cages, it is necessary for the operator to wear gloves to avoid possible pecking by the animal

charge of restraining the cockerel wears gloves of resistant material (Fig. 4.3). It is not advisable to get too close to the cages as it is possible to receive a painful peck from one of the bird while carrying out the procedure.

Once the rooster is immobilised, it is removed from the battery cage and taken to the insemination rack. Here it is immobilised by one of the operators by the legs and wings, and then the other operator starts to perform the stimulation and semen collection operations (Fig. 4.4).

The Burrows and Quinn technique (Burrows and Quinn 1937, 1939) for semen collection is carried out by lightly massaging the terminal dorsal area above the cloaca and in the area immediately above the tail. The hand is placed around the cloaca, and the male's penis is visualised. It is a rudimentary organ, which can vary in size, formed of a folded mucous membrane with a central groove through which the ejaculate of the guinea fowl will slide. By stimulating the area around the cloaca, with the penis on the outside, tumescence is achieved. This procedure is carried out with the index finger and thumb of one hand and with the other the container, a petri dish, a test tube, a suction tube, etc., in which the semen will be collected is held. Once the cockerel's penis is swollen and everted through the animal's cloaca, a gentle massage is performed on the lateral area of the penis, exerting gentle pressure on the lateral bulbs in order to "milk" the semen and thus take advantage of the maximum amount of ejaculate. The milky white sperm flows through the penile groove (Fig. 4.5). In guinea fowl, the volume collected is around 50–150 μl per rooster. The concentration of spermatozoa is very high on average, reaching 7–8 million spermatozoa/μl.

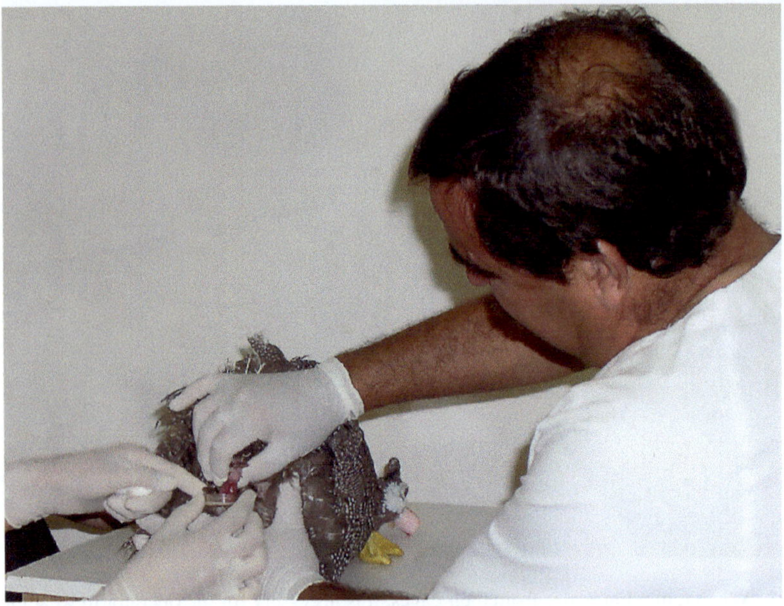

Fig. 4.4 Handling of the male to extract semen requires the use of a raised table or "rack" on which the bird is supported

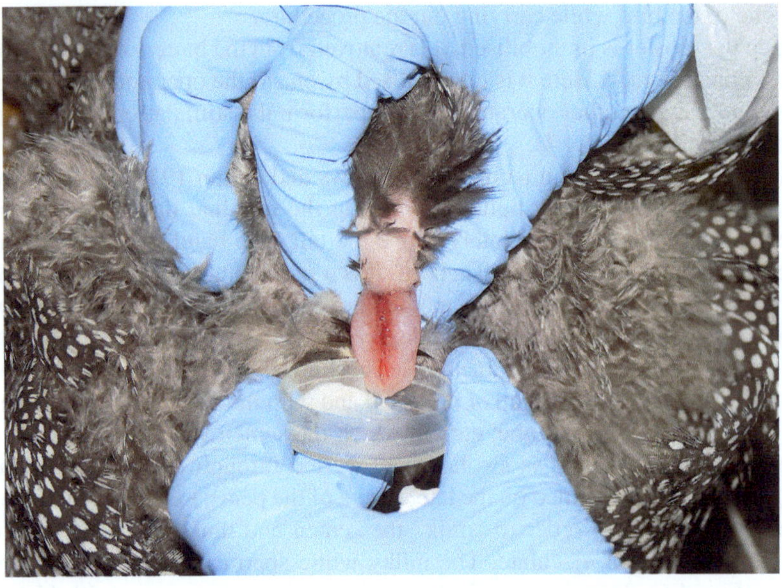

Fig. 4.5 Collection of semen from guinea fowl

All these operations should be carried out as gently as possible, as the mucosa of the rooster's penis is very sensitive and can be damaged, with negative repercussions on the bird's reproductive performance and the quality of the semen collected. Semen should be used within 10–15 min after collection operations.

A recommended action before starting semen collection and insemination in females is to pluck the ischiorectal fossa in order to obtain clean semen samples and to facilitate insemination in hens.

The frequency of sperm collection allows the total number of sperm produced per cockerel to be modified. The change from one to two sperm collections per week (two inseminations per week) increases the number of sperm produced by each cockerel during the week by almost 50%.

At the end of the rooster's reproductive life, there is often a sudden decrease in the number of sperm collected per ejaculate. This decrease is due to a drop in testicular production. In this case, it is recommended to collect semen from the guinea fowl every 5 days instead of twice a week.

Insemination Dosage in Guinea Fowl

According to several authors, it is necessary to dilute the semen to improve the insemination results. However, inseminations with pure semen carried out at the La Orden-Valdesequera Research Centre have proven to be effective, achieving the same or even better results than when diluted semen is used.

During the first half of the laying period, it is recommended to use insemination doses containing between 80 and 90 million spermatozoa, which is achieved with approximately 8–10 μl/female; in the second half of the reproductive period of the females, the doses should be increased to 110–120 million (approximately 20–40 μl/female). This increase in the insemination dose is due to the fact that, with age, the semen storage capacity of the hens decreases, so this must be taken into account in order to maintain fertility levels.

To obtain such quantities of semen, approximately one male for every four to six females is needed.

Insemination of Females

Insemination of the hens starts once the semen has been collected from about six cocks.

According to studies carried out on French farms, three people can inseminate 2500–3000 guinea fowl per day and carry out sperm collection.

The insemination technique is the one proposed by Burrows and Quinn (1937, 1939), via the genital route. To perform this technique, only the rear of the female is removed from the battery cage. The female is then held by the dorsum with the left hand while the tail is moved forward; the folds of the cloaca are then pulled taut with the right hand and the index finger and thumb, while the abdomen is pressed

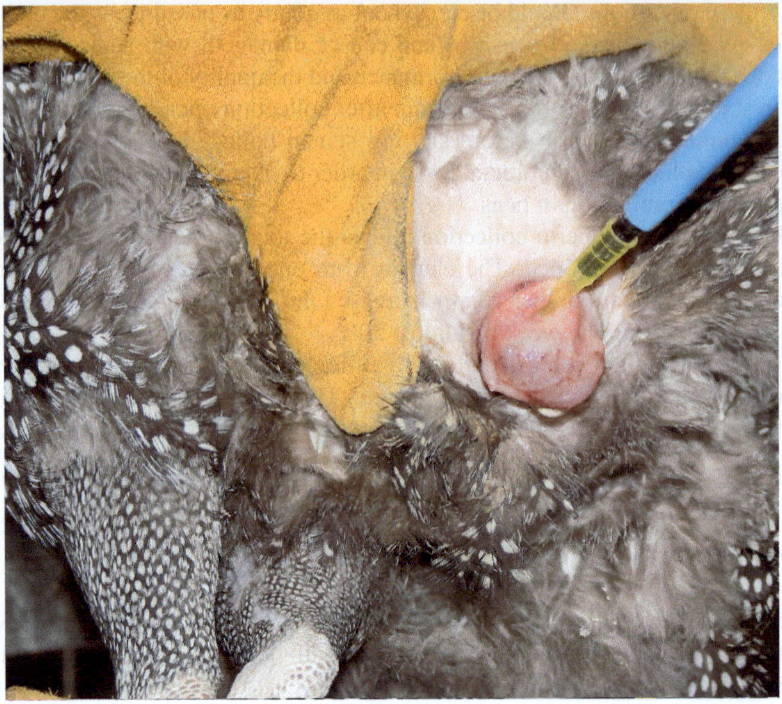

Fig. 4.6 Introduction of semen into the female using a micropipette. Once the vagina has been everted, the semen is deposited to a depth of between 1 and 3 cm

forward until the oviduct emerges through the cloaca. At this point, the semen dose is deposited with the tip of the micropipette (Fig. 4.6).

The best results are obtained by performing inseminations at a depth of about 1–3 cm from the entrance of the vagina. Insemination too deep into the vagina increases the health risk and trauma to the female genital tract.

An insemination that is too superficial may cause the sperm to be expelled, due to the secretions that exit through the oviduct, which can sometimes be very abundant and literally sweep the insemination dose outside.

After insemination, abdominal pressure shall be released, and the hen shall be returned to her cage.

The spermatozoa, after being introduced, will migrate to the infundibulum, where they will fertilise the egg. A small amount of sperm will be stored in the utero-vaginal junction, where they will remain for an as yet undetermined period of time.

When choosing the time of insemination, it should be kept in mind that the hatching rate varies considerably, depending on the time between oviposition and insemination. This rate is optimal for inseminations carried out at least 3 h before or after oviposition.

Laying times are synchronised with the on/off light cycle. According to several authors, 50% of ovipositions take place 7–9 h after lights are turned on, when daily lighting is administered as a single 14-h light period. It is therefore advisable to carry out inseminations within 4 h of switching on or within 2 h of switching off the light. In any case, it is necessary to plan inseminations according to the cycle of lightness and darkness applied on the farm.

Material to Be Used During Insemination

The instruments used for insemination can vary widely.

A small diameter petri dish can be used to collect the ejaculate from the bird in order to be able to evaluate it individually before mixing it with the semen from the other cocks to be used in that insemination batch. The semen contained in the dish is aspirated by means of a plastic Pasteur pipette with a bulb. Once the ejaculate from approximately six cocks has been collected, the Pasteur pipette bulb is placed in a test tube, cutting it in half to allow easy access to the ejaculate.

A micropipette is used to access the semen in the bulb and take the necessary amount for immediate insemination. In the case of the plastic tips of the micropipette, it is advisable to trim the distal end portion (approximately 1–2 mm), to avoid scraping the mucosa of the hens' vagina during insemination.

It is essential for the operators to use protective clothing. Handlers must use protective goggles, gloves made of resistant material for capturing and immobilising the guinea fowl (latex gloves for the semen handler) as well as protective masks and suitable clothing and footwear.

It is necessary to observe the rules of hygiene and disinfection of equipment and personnel before and after insemination operations (Pérez 1966; Hoffman and Völker 1969; Sauveur and Reviers 1992; Hafez 1993).

Semen Characteristics

Spermatozoa Morphology in Guinea Fowl

The morphology of avian spermatozoa (Fig. 4.7) is different from that of mammals. In birds, spermatozoa are generally elongated in shape with a slightly twisted head and a large tail.

Spermatozoa are formed within the seminiferous tubules of the testes; once formed, they consist of a flattened head containing the nucleus, a tail containing the apparatus necessary for cell motility and a neck linking the head to the tail.

Avian spermatozoa are filiform. The head is morphologically simple, the acrosome is cone shaped, contains homogeneous material and is enveloped by a continuous acrosomal membrane. There is no equatorial segment and no dense post-acrosomal lamina, typical of mammals. The acrosome is separated from the nucleus by the sub-acrosomal space occupied by the perforatorium.

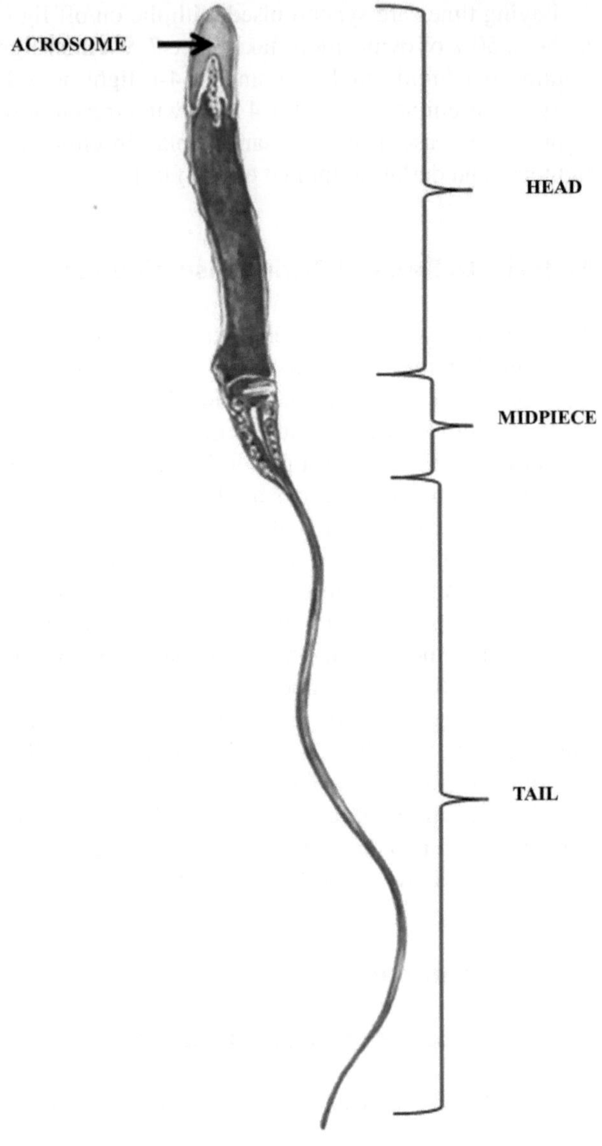

Fig. 4.7 Structure of a poultry spermatozoon. (Original graphite drawing by Diego Del Moral Martínez)

The nucleus contains condensed chromatin granules; it is enveloped by a double nuclear membrane, and the neck region linking the head to the caudal segment is formed by a centriolar complex. The axoneme is the motor component of the tail; it originates from the distal centriole and is surrounded by about 30 mitochondria forming the medial segment. The dense cytoplasmic ring marks the distal border of the medial segment and the proximal border of the main segment; however, it lacks the fibrous sheath and dense outer fibres seen in mammals, but the axoneme is enveloped in an amorphous sheath.

The overall structure of these spermatozoa is common in all domestic fowl species. The filiform appearance of the nucleus and the weak development of the acrosome are common characteristics of the spermatozoa of gallinaceous birds Anatidae, Anseridae and *Meleagris*.

In addition to spermatozoa, guinea fowl semen contains a seminal plasma consisting of the non-gametic portion of an ejaculate. It is a nutrient medium for spermatozoa and their transport vehicle within the female reproductive tract. It contains between 50 and 60 chemical components, and it is low in chloride, citrate, ergothioneine, inositol, phosphorylcholine and glycerophosphorylcholine, but the main anion is glutamate.

Assessment of Guinea Fowl Semen

Macroscopic Estimation of Guinea Fowl Semen

The first evaluation of the ejaculate of the male guinea fowl must be done in the collection pipette reservoir. The quality of the semen in this first evaluation phase is greatly influenced by the collection technique used and the care of the operator during collection. This avoids external contamination and ensures that the highest possible volume of ejaculate is obtained from each guinea fowl cock, without damaging the reproductive tract of the rooster and limiting its performance.

Semen Quality

Good semen collection technique is essential to optimise the results of artificial insemination. For this, it is necessary to handle the instruments and the cocks carefully and hygienically so as not to contaminate the semen, given that in case of contamination with remains of faeces, red blood cells or excessive epithelial cells, due to a bad collection technique, anomalies can occur, such as sperm aggregation phenomena, which consist of the adherence of the sperm membranes to other cells present in the semen (epithelial cells, red blood cells, detritus, etc.) and that lead to a decrease in fertility.

First of all, the volume of semen obtained shall be assessed. In the case of guinea fowl, the volume ejaculated by one animal ranges between 50 and 150 μl; this volume can be directly visualised by the graduations on the semen collection device used. In guinea fowl, the volume of ejaculate depends on many factors: age (decreases with age), health status, feeding and skill of the collector.

The colour of the semen can range from clear white to slightly yellowish and slightly creamy in consistency (Fig. 4.8).

Fig. 4.8 Good semen collection technique has a significant influence on semen quality

Abnormalities in the Colour and Consistency of the Semen

- Reddish colour: this can be caused by scraping and bleeding of the penile mucosa due to poor stimulation technique during ejaculate extraction. In this case, blood mixes with the sperm and contaminates it, reducing the quality of the semen.
- Brown colour: this is caused by contamination of the semen with faeces from the bird or by systemic diseases; in the latter case, the affected semen must be discarded for inseminations. To avoid contamination of semen with faeces, we must remove the feed from the cockerels 8–12 h before collection.
- Semen that is too liquid in consistency: this should also be rejected for insemination.
- Semen with foreign bodies: this is often caused by not conditioning the genital area of the male properly; to avoid this, we must remove all the feathers from the area around the cloaca, both in males and females. In the case of cockerels, we will be able to reduce, to a great extent, the contamination of the semen by feather debris, skin flakes and dirt, amongst others.

Other Parameters to Be Taken into Account in the Macroscopic Evaluation of Semen

- Macroscopic mass motility: assessed by observing the collection tube and detecting whether mass movement or swirling is present. The test is considered as positive or negative.
- pH: assessed by removing a drop of semen from the tube and placing it on a pH indicator strip. Samples with a high concentration of spermatozoa and motility tend to have a more acidic pH than contaminated samples; if they have a high number of dead spermatozoa, they tend to have a higher pH. The pH strip should not be inserted into the tube so as not to alter the semen with the reagent in the strip.
- Schalm test: performed to detect the presence of leukocytes, but only in cases where the presence of leukocytes is suspected.

Microscopic Estimation of Guinea Fowl Semen

The semen characteristics to be taken into account in its microscopic evaluation are motility (mass and individual), vigour, sperm concentration and sperm morphology.

Motility and morphology can be assessed using a microscope; before doing this, it is necessary to dilute the semen in order to observe the movements of the spermatozoa more clearly. This evaluation should be carried out as soon as possible so that the semen does not lose its optimal conditions for fertilisation.

To determine motility (Mi.M.M, "Microscopic Mass Motility"), a drop of pure semen is placed on a microscope slide at 36–37 °C on a heated microscope stage and observed at 40× magnification (stereomicroscope), evaluating the presence of omega waves. It should be evaluated near the edge of the drop, where the depth of the drop is shallower and easier to observe. The scale used is from 1 to 5, with 1 being a semen with no waves and 5 when the waves are moving rapidly, forming whirlpools. Within these parameters, intermediate points are considered. A minimum acceptance value of 3 Mi.M.M. is considered.

Individual motility (IM) is determined by diluting the semen in sodium citrate. A thick drop of semen is placed in a tube with about 2 ml of the citrate solution, which should be at the same temperature as the semen, already in the water bath. Once the semen has been diluted, a drop of the dilution is removed and placed on a slide at a temperature of 36–37 °C, and a coverslip is placed on top, also at the same temperature. The dilution is observed under the microscope, if possible on a heated stage. A field is examined, and the spermatozoa that move in a progressive, straight-line manner are subjectively assessed, as these are the ones that cross the field of observation. Spermatozoa that rotate in a circle or move in an oscillatory manner are considered to have abnormal movements. The percentage indicated is the percentage of spermatozoa with progressive rectilinear movement out of the total number of spermatozoa accepted, with the minimum acceptable value being 50%.

The classification of semen according to individual motility is as follows:

VG (very good) = 80%–100% of mobile cells
G (good) = 60%–79%
R (regular) = 40%–59%
P (poor) = <40%

Vigour is assessed at the same time as IM, taking into account the speed with which these sperm move through the field. The scale used is from 0 to 4, with immobile spermatozoa being rated as 0 and spermatozoa that move quickly through the field and are difficult to follow visually being rated as 4. Within these parameters, intermediate points are considered. A vigour of 3 is considered as a minimum acceptable value.

The sperm concentration of an ejaculate is the determination of the number of spermatozoa per unit volume, as it is essential for seminal analysis—seminal doses are prepared according to the volume concentration. The determination of the concentration is done using counting chambers, such as the Bûrker or Neubauer chamber, and a red blood cell counter, where the dilution is prepared with formalin citrate. In addition to being used to calculate the sperm concentration, this chamber is used to count morphological abnormalities, both in the head and tail of spermatozoa.

The assessment of the morphology (vital staining, morphology sensu stricto and acrosomy) of the spermatozoa is done by observing microscopic preparations or smears. These can be used to assess whether the observed morphology is adequate or if there are foreign elements in the preparation, which can detect fertility problems when the percentage of abnormal forms is high (Figs. 4.9, 4.10, 4.11, and 4.12) (Bakst 1980; Nwakalor et al. 1988; García-Herreros 2016; Korankye et al. 2020).

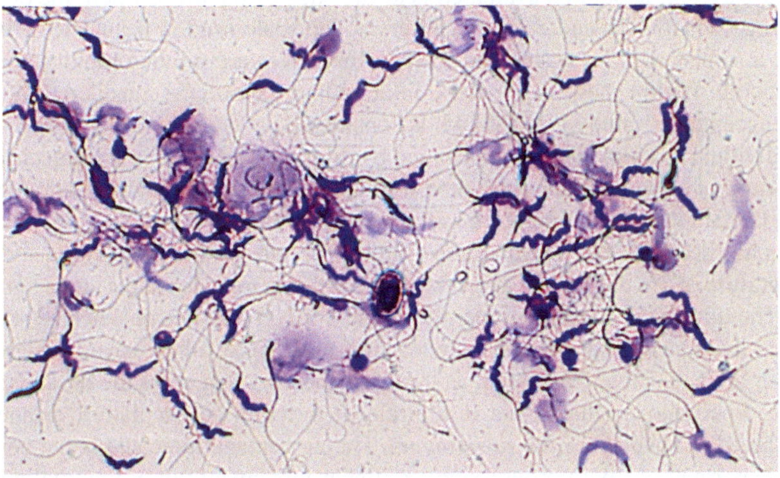

Fig. 4.9 Guinea fowl sperm aggregated around epithelial cells and a red blood cell; note also the presence of abnormal shapes (rapid panoptic staining)

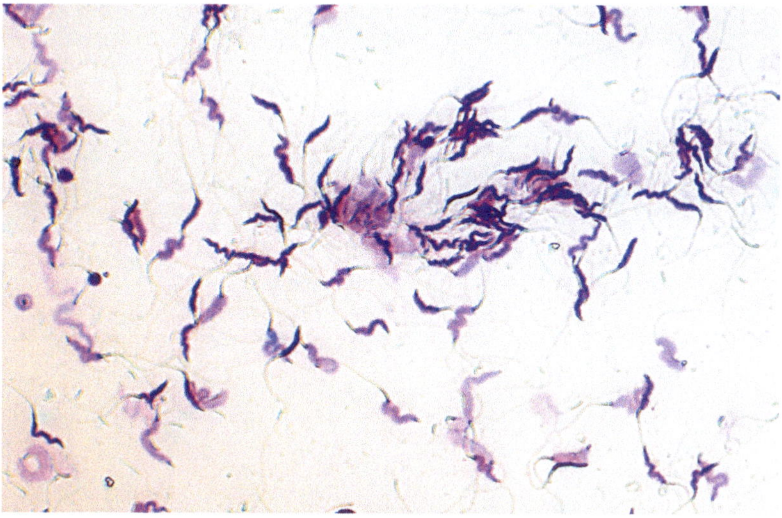

Fig. 4.10 Aggregation of spermatozoa around epithelial cells (rapid panopticon)

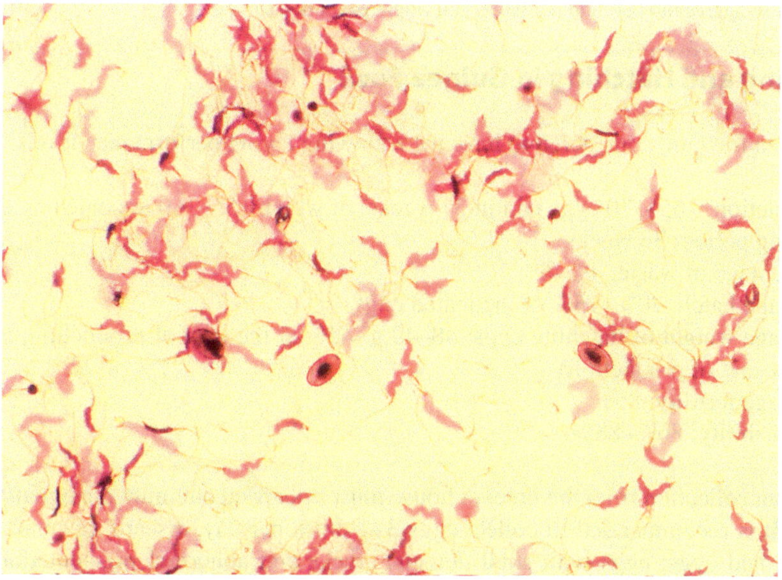

Fig. 4.11 Presence of blood (red blood cells) in guinea fowl sperm (rapid panopticon)

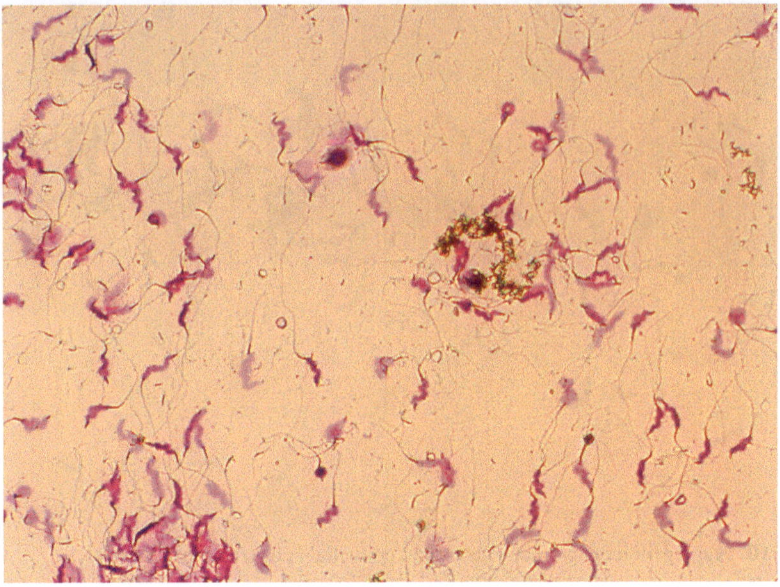

Fig. 4.12 Aggregation of spermatozoa around traces of dirt present in the semen of guinea fowl; note also aggregation around red blood cell (rapid panopticon)

Laying and Hatching in Guinea Fowl

Guinea fowl are considered to have the following laying performance levels:

Production: 165–170 eggs laid in 37 weeks, of which 155–160 are hatching eggs.
Laying period: 36 weeks
Percentage of waste: 4%–5%
Average hatch: 70%–75% of incubated eggs
Average weight of hatching eggs: 48–49 g (45 g at the start of production, 50 g at the end of production)
Average fertility: 85%
Hatchability: 86%–88%

Egg collection from the breeder house must be carried out immediately after laying. It is recommended to collect eggs 3–4 times per day, discarding small, dirty, deformed eggs; eggs with shell defects, cracks and abnormal pigmentation; and eggs with excessively porous shells. The selected eggs shall be disinfected within 2 h of collection using the following formula:

- 40 ml formaldehyde solution
- 20 g potassium permanganate
- 40 ml of water

Fig. 4.13 Guinea fowl eggs arranged with the acute pole downwards for hatching

The disinfection procedure consists of placing the eggs in a closed room in which an open container (dish or pan) with the described solution is placed for 20–30 min.

After disinfection of the eggs, they are stored in a chamber under the following conditions of temperature, humidity and time:

- For short storage periods (up to 1 week): 17–20 °C/85% humidity
- For long storage (up to 15 days): 12.5–15 °C/85% humidity

Before loading the eggs into the incubator, it is advisable to pre-warm the eggs by storing them in a room at a temperature of 22–24 °C for 8–12 h before loading the incubator. Once the selected eggs have been loaded into the incubator (Fig. 4.13), the incubator shall be programmed as follows:

- Incubation temperature: 37.4–37.6 °C, when eggs have been stored for a short time ($\approx$ 1 week)
- Incubation temperature: 37.7 °C, when the eggs have been stored for about 2 weeks
- Relative humidity: 60%
- Turning: every hour; turning simulates natural incubation and avoids adhesion of the embryo to the eggshell

Eggs can be candled after 10–12 days of incubation to remove defective, broken, clear and dead embryos.

After 24 days of incubation, eggs are transferred to the hatcher, which works in a similar way to the incubator, although in this case there is no turning; from the alveoli of the incubator, the eggs are transferred to trays with grids, where they are kept from day 24 until they hatch on day 28 approximately.

The reference parameters in the hatchery are as follows:

- Temperature: 37–37.5 °C. We lower the temperature to avoid overheating of the guinea fowl during hatching.
- Relative humidity: 80%. We increase humidity to facilitate hatching.

Hatching Egg Health Programme

From laying to hatching, the guinea fowl egg undergoes a period of time in which the microbial load on the shell can cause sanitary problems if conditions are suitable for germ proliferation. To avoid this, it is necessary to establish a sanitary management programme for hatching eggs.

Shell germs can be transmitted to the egg vertically, from hen to egg, or horizontally from a contaminated egg to a nearby egg at laying, collection or storage. In both cases, the contaminations that occur lead to various health problems, such as disease transmission between flocks, hatching and hatchling problems as well as spread of infectious processes on the production farm, with consequent economic losses and risk for the consumers of the final product (Krunt et al. 2021).

According to a different legislation, the health programme should be structured around the following points:

- Egg collection. The production of eggs in batteries improves their microbiological quality, as the eggs remain clean until they are removed by the operator. Egg collection should be carried out as many times as possible throughout the day, with a minimum of twice a day being recommended. When removing the eggs, they should be placed in a clean and disinfected container, free of remains of broken eggs from other collections and dirt.
- Eggs that are found to be defective, non-hatchable, must be rejected. This is the case for eggs with shell abnormalities such as excessive porosity, cracks or breaks, strange colouring, abnormal size (too large or too small), deformities, double yolks, etc.
- Dirty eggs, contaminated by hen droppings, will also be rejected for incubation—it has been shown that incubated dirty eggs can result in up to 20% fewer hatchings, leading to up to 15% chick mortality in the first week.
- Handling of hatching eggs. Eggs should be collected from the batteries as often as possible and disinfected and refrigerated as soon as possible to prevent bacteria that are in a position to multiply from entering the egg through the pores of the shell into the egg.
- Disinfection of hatching eggs. Disinfection shall be carried out as soon as possible after collection and before storage as described above.

Fumigation personnel must be adequately trained, and the products used must be authorised by current legislation. The rooms where fumigation operations are carried out must be equipped with adequate ventilation systems for the elimination of toxic gases.

Once the eggs have been disinfected, they should be stored refrigerated under appropriate conditions as soon as possible, at the temperatures and humidities described above. It should be noted that for both storage and incubation, the eggs should be placed in the alveoli with the acute pole downwards.

In order to avoid excessive "sweating of the eggs"—a phenomenon produced by the condensation of humidity on the surface of the eggs, due to the sudden contrast of temperature and humidity when they leave the storage chamber—and that the viability of the embryos may be affected, they should be transferred to a pre-warming chamber (21–23 °C, the afternoon before their entry into the incubator). For storage times of up to 7 days, they should be kept at these temperatures for 12 h, and for storage times of up to 14 days, the same temperature should be used, but during the 18 h prior to placing them into the incubator.

Incubation

The eggs shall be placed in the alveoli of the incubator (Fig. 4.14), taking care to place the eggs that have been stored the longest first, in the case of long storage periods. The incubator shall have a properly functioning ventilation and turning system to ensure air renewal inside the incubator and to avoid embryo adhesion to the egg membranes.

Fig. 4.14 Industrial incubator loaded with guinea fowl eggs

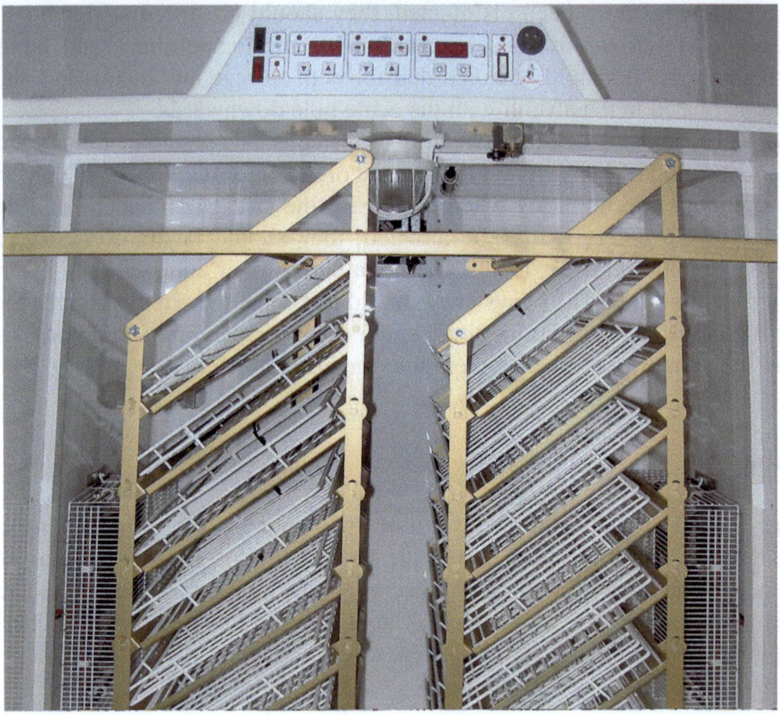

Fig. 4.15 Incubator model with automatic turning system

To improve the performance of the incubator (Fig. 4.15) and thus increase the number of hatches, especially in winter incubations, the air in the incubation room should be tempered to around 23–25 °C. In case of summer incubations, the air in the room should be cooled as much as possible to this temperature.

The frequency of turning is programmed directly in the incubator; it is a very important function to avoid the adhesion of the embryos of the keets to the internal membranes of the egg. In the early stages of incubation, the higher the frequency of daily turning, the higher the hatchability of the eggs.

After 24 days of incubation, the eggs should be transferred from the incubator to the hatcher (Fig. 4.16), which should be brought to the proper temperature hours before. During this operation, the eggs can be inspected, and those with any breakage, that are clear or that have aborted in the first days of incubation can be eliminated. In the case of the hatcher, turning the eggs is not necessary. The eggs will remain inside until the 28th day of incubation.

It is very advisable, when placing the eggs in the hatcher drawers, to place them as close together as possible, leaving the minimum space between them. so that the keets, at hatching, are close together, which helps prevent lead weakness—a condition in which the chicks' legs are splayed, a defect which limits their mobility throughout their lives. It is also advisable, to circumvent this problem, to line the

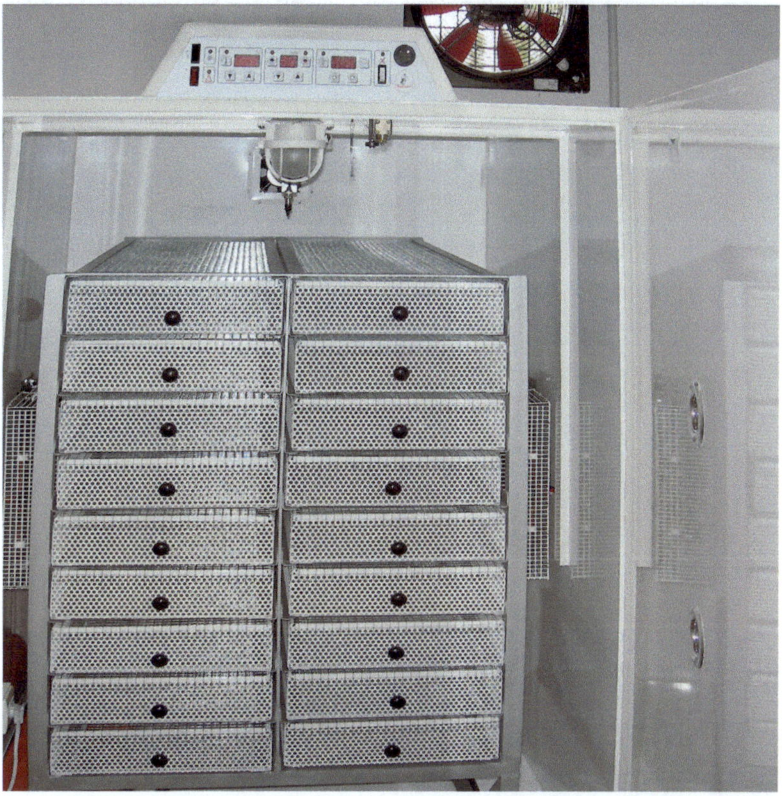

Fig. 4.16 Hatcher with latticed drawers to prevent guinea fowl from escaping

floor of the hatching trays with clean paper, thus helping the guinea fowl to better hold on to the hatch and not to slip.

The following guideline parameters can be used to determine when to remove the guinea fowl from inside the hatcher:

- When the hatching, more than 60% of the keets must be wet on the head and neck (Fig. 4.17).
- Thirty-six hours before removal, about 10% of the chicks should have hatched.

Of the hatched chicks (Fig. 4.18), it is necessary to eliminate those with deformities, limb problems, dehydration, unhealed navel or low weight (the average weight is 32–34 g).

This "triage" is necessary to obtain homogeneous and problem-free flocks and will consist of eliminating all birds with defects, such as underweight, malformations, deformed legs or crooked toes, deformed beaks, poorly healed navels and lack of vitality.

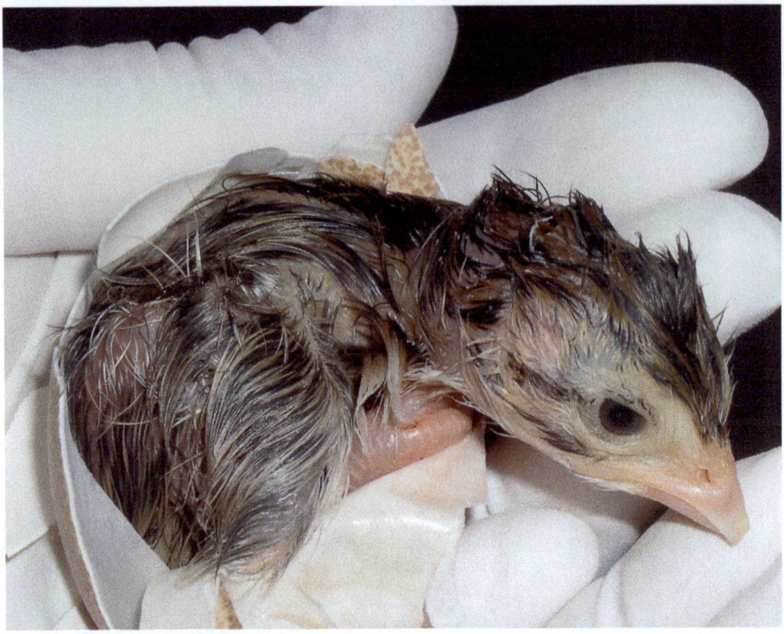

Fig. 4.17 Chick leaving egg with wet head

Fig. 4.18 Keet ready to be sent to the breeding park

Fig. 4.19 Box of guinea fowl fresh from the hatchery and ready for transport to the breeding park

The remains of incubation (eggshells, membranes, remains of chicks, etc.) must be deposited in leak-proof containers for subsequent disposal by authorised procedures. (*A fundamental aspect of the incubation process is the hygiene of the hatchery room, which must be washed daily with water mixed with bleach and a chlorinated detergent, a practice which must also be applied to the egg storage chamber.*)

Once the chicks have been removed from the hatcher, and the "triage" has been carried out, they are transferred to the rearing pens in plastic trays (Fig. 4.19).

Breeding rooms should also be tempered and conditioned hours before the birth of the guinea fowl to facilitate acclimatisation to their new environment (Muriel 2005; Muriel and Serrano 2006, 2007).

Post-hatchery Operations

After unloading the incubators or hatchers, they shall be cleaned and disinfected by removing all organic debris from inside the machines and cleaning them inside and outside with pressurised chlorinated water, followed by the use of a contact disinfectant.

To identify hatching abnormalities in the chick, unhatched eggs at the end of incubation should be examined and classified into the following categories:

- Infertile eggs: to distinguish infertile eggs from aborted eggs in the first hours of life, the inside of the egg shall be examined, and we shall see, in the case of the infertile egg, the embryonic disc of an intense white colour, and in the case of the aborted egg, the embryonic disc with a clear space in the centre.

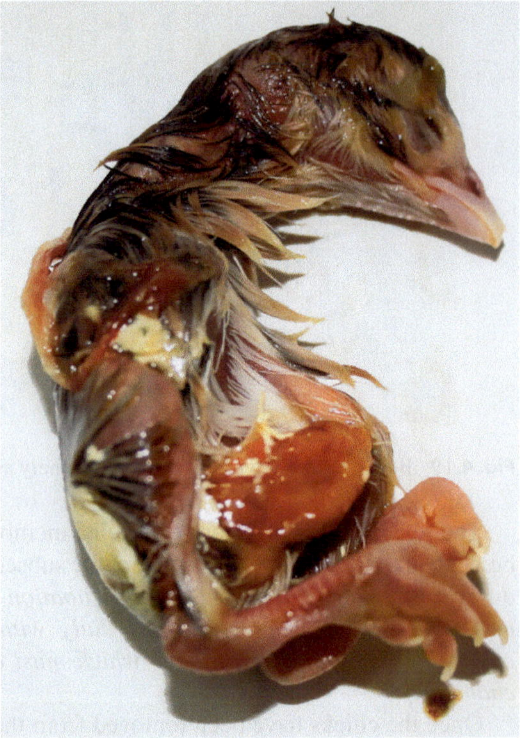

Fig. 4.20 Keet fourth-stage abortion

- First-stage abortions (0–4 days): causes of abortion in the first hours are due to anomalies in the storage chamber, such as extreme temperatures (too cold or too high) or anomalies in the incubator.
- Abortions in the second stage (5–17 days): this second stage is marked by the appearance of the eye in the embryo, and the frequent causes of death in this period are high temperatures at the beginning of incubation, inadequate turning, eggs contaminated with encephalomyelitis and salmonellosis and lack of vitamin E in the diet of the breeders.
- Abortions in the third stage (>18 days) (Fig. 4.20): these include mortality in the incubator and hatcher; causes of infertility in this stage are incorrect humidity and temperature in this period, turning errors, lack of ventilation, contaminated eggs, lack of micronutrients in the breeder's diet, mycoplasmosis, Newcastle disease and infectious bronchitis.

On examination of newly hatched chicks, various symptoms related to several other aetiologies may appear (Table 4.1) (Offiong and Abed 1980; Beard and Hanson 1984; Woernle 1994, MAPA 2004; Jegede et al. 2024).

Table 4.1 Possible anomalies which, according to Professor Castelló (1991, 1996), may arise in the evolution of eggs during incubation

Symptoms	Aetiology
Keets too small	Underweight eggs Thin, porous shells Low humidity in incubator High temperature in hatchery
Weak keets	High temperature in hatchery Delay in the removal of the Guinea fowl Nutritional deficiencies in breeders Lack of ventilation
Keets with shells stuck on	Low humidity in hatchers Temperatures too high at birth
Keets with swollen belly	Overweight eggs Omphalitis High humidity in incubators Poor ventilation in hatcheries Low incubator temperatures
Sticky keets	Humidity too high at birth Inadequate or poor air circulation in hatchers Low temperatures at birth Old eggs
Deformed keets	Tarsal dislocation: sliding birth trays, reproductive nutritional deficiencies Crossed beak: genetic defect or viral infection in breeding females Crooked neck: Mn deficiency in breeding females Short legs: genetic problem or deficiency in breeding females Red hocks: old eggs or vitamin B deficiency Eye abnormalities: Genetic problems and overheating

Raising Guinea Fowl Chicks

The newly hatched guinea fowl are housed in the breeding pens, inside a galvanised sheet metal enclosure with heat, food dishes and drinking troughs (Fig. 4.21). The pens must be conditioned 48 h beforehand, and a sanitary break of at least 14 days must be carried out.

The fences within the rearing pens shall be partially opened from the first week of age and completely removed from the fourth week onwards. The guinea fowl shall remain freely in the rearing pens until the sixth to seventh week.

An important aspect of rearing is temperature control during the first weeks (Table 4.2).

Another aspect to consider is that during the first 3 days of the guinea fowl's life, the lights should be on 24 h a day so that the birds become familiar with the elements of the park, feeders, drinkers, etc., in order to be able to establish the most suitable light-dark cycle later.

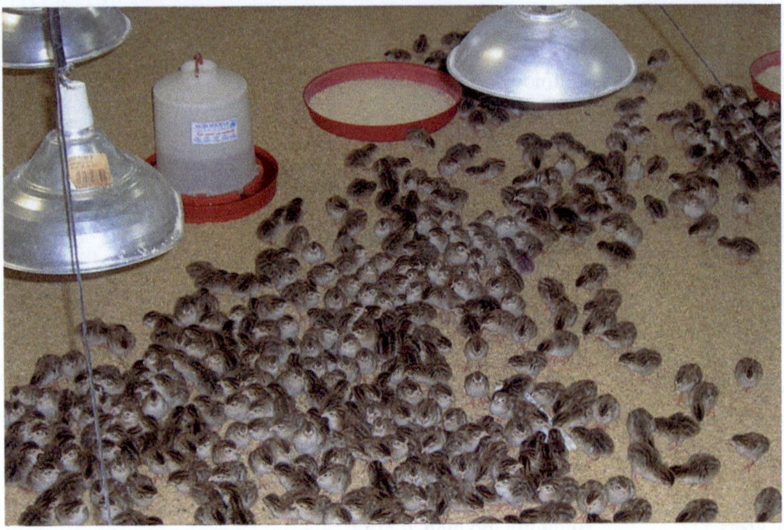

Fig. 4.21 Keets housed in their breeding park

Table 4.2 Recommended temperatures for breeding guinea fowl in France, according to Dr. Le Coz (1992)

Age of the guinea fowl	Temperature under the spotlight (°C)	Ambient chamber temperature (°C)
First 3 days	36–38	27–29
Seventh day	36–38	27–29
Second week	36–38	26–28
Third week	34–36	25–27
Fourth week	27–29	25–27
Fifth week	24–26	25
Sixth week	24–26	20–24

Fig. 4.22 Guinea fowl in open-air parks with plenty of space at their disposal

Rearing of the Guinea Fowl

In Extremadura, where the farming system is extensive, rearing starts from the sixth to seventh week and lasts until the guinea fowl are 14 weeks old. In France, where the system is predominantly intensive, they are usually slaughtered at 12 weeks.

In the extensive system, the birds are moved outdoors to the open pens to be reared (each bird has >1 m^2), where they can walk and fly freely (Fig. 4.22). It is very important to consider the weather conditions during the first days of transfer from the brooding closures (covered) to the rearing paddocks (open); in cold or rainy seasons, the change should be made very gradually so that the birds can acclimatise, removing the heaters little by little, until they are all gone.

In the parks, the feeders are replaced by feed hoppers, and the starter drinking troughs are replaced by automatic drinking troughs.

Collection, Transport and Disposal of Guinea Fowl

Both the catching of the birds and the transport to the slaughterhouse are very important stages in a guinea fowl farm. Guinea fowl have a very nervous nature and are therefore very sensitive to stress, a factor that must be mitigated as much as possible, allowing for an increase in carcass quality and a decrease in economic losses. One of the main factors that decrease carcass quality during transport is the shocks the birds may suffer during catching, transport and unloading at the slaughterhouse. Another factor to take into account is the high number of deaths that can occur in a batch due to poor management of transport to the slaughterhouse; in the case of guinea fowl, this circumstance can be dramatic, which will result in high economic losses.

Capture and Caging of Birds

Before catching the guinea fowl, it is necessary to fast them from feed for 12 h prior to loading and transport. Only the feed shall be withdrawn, leaving water freely available to avoid dehydration of the birds, especially in summer.

In order to capture the guinea fowl and immediately introduce them into the transport cages, several points must be taken into account, which are fundamental tin reducing the stress of the birds:

- The guinea fowl will be captured manually.
- During the capture of the birds, the greatest possible silence should be maintained, without loud noises, shouting or sudden movements.
- In order to facilitate movement within the paddocks, it is advisable to remove any foreign elements, such as feeders, hoppers, automatic drinkers, etc., which will hinder the movement of the operators and the birds.
- Birds should be captured in semi-darkness to reduce stress.

- Avoid crowding birds in corners and thus reduce deaths by suffocation.
- It is useful to use partition systems which will help keep animals in the corners of the parks, but excessive numbers of birds should be avoided, as crowding and suffocation can occur.
- If the bedding is clean prior to catching the birds, excessive dust build-up in the enclosures can be avoided, which would otherwise make it difficult to catch the birds; to avoid this, the extractors in the paddocks should be switched on during catching.
- Pick up guinea fowl by the hind legs and never by the neck or wings and never more than three or four at a time in each hand; handling should be as gentle as possible.
- A reasonable number of birds should be placed in the cages to avoid suffocation or hyperthermia in hot weather; in the latter case, the number of birds per cage should be reduced.

Transport

The transport of the guinea fowl, once captured and caged, shall be carried out by means of the livestock transport lorry properly documented for this purpose.

It is important to drive smoothly and steadily to the slaughterhouse, and stops are not advisable, as while the vehicle is moving, the bird is cooled, there are no problems of hyperthermia and stress is reduced.

In hot weather, it is advisable to load and transport during the night, taking advantage of the cooler hours of the day.

Regulations established by the relevant authorities for the transport of live animals must be strictly followed.

Unloading

The birds must be unloaded at the slaughterhouse from the unloading dock, where the operators open the transport cages and hang the guinea fowl on the conveyor belt for stunning. The operators must be aware that these birds are more vigorous and more rustic than traditional meat chickens; they often resist handling and may even bite hard at the personnel in charge of these functions in the slaughterhouse, being slippery and being able to escape, causing additional losses during transport which are not at all desirable.

Transport cages must be washed and disinfected before re-use.

Stun and Slaughter

Birds freshly unloaded from the cages are hung by their legs on the hooks of the transport chain; guinea fowl, when hung, usually remain still, although flapping may occur, and are then moved to the stunning area.

The birds in the stunning area pass through a channel filled with water, through which an electric current is passed; the birds' transport chain descends until the head and neck of the guinea fowl are introduced into the water, at which point they are either stunned or killed, depending on the intensity and voltage applied.

Slaughter is carried out by cutting the veins and arteries in the neck; depending on whether a carcass with or without head is to be obtained, the technique may vary.

The bleed-off of the guinea fowl is carried out in the bleed-off channel.

Afterwards, the typical operations in any poultry slaughterhouse are carried out: scalding, plucking and de-feathering, finishing, evisceration, pre-cooling, sorting, cooling and storage.

In the case of the guinea fowl, it is of special interest for a good presentation of the carcasses to carry out a thorough plucking, trimming and finishing, as the density of feathers of this species is higher than that of chickens for meat, and there may be remains of quills and feathers that diminish the quality of the carcass.

Once the guinea fowl have been slaughtered, weighed and classified, the carcasses are taken in boxes to standing refrigeration chambers or tunnels, at a temperature of around 4 °C, in order to cool them and allow them to mature; this operation lasts several hours, after which they are ready to be sent to market (Castelló 1991, 1996; Le Coz 1992; Jouffroy D'Abbans 1996; Cooper and Adabi 2012).

Competing Interests The authors have no conflicts of interest to declare that are relevant to the content of this chapter.

References

Bakst MR (1980) Vitrasturecture of chicken and Turkey gametes, fertilization, and oviductal sperm transport: a review. In: Eighth international congress on animal reproduction and A.I., Madrid, vol 2, pp 511–517

Beard CW, Hanson RP (1984) Newcastle diseases. In: Hofstad MS, Barnes HJ, Calnek BW, Reid WR, Yoder HW Jr (eds) Diseases of poultry, 8th edn, pp 452–470

Burrows WH, Quinn JP (1937) The collection of spermatozoa from the domestic fowl and Turkey. Poult Sci 16(1):19–24

Burrows WH, Quinn JP (1939) Artificial insemination of chickens and turkeys (No. 525). US Department of Agriculture

Castelló JA (1991) Chicken meat production. Real Escuela de Avicultura, Barcelona

Castelló JA (1996) Exploitation of guinea fowl. In: Buxadé C (ed) Zootecnia. Bases de Producción Animal, 10th edn, pp 282–298

Cooper RG, Adabi SG (2012) Guinea fowl production. Lulu Enterprises, London

García-Herreros M (2016) Sperm subpopulations in avian species: a comparative study between the rooster (*Gallus domesticus*) and Guinea fowl (*Numida meleagris*). Asian J Androl 18(6):889–894

Hafez E (1993) Reproduction and artificial insemination in animals, 6th edn. Interamericana, Mexico, p 373–375, 398–411

Hoffman G, Völker H (1969) Anatomy and physiology of domestic fowl. Acribia, Zaragoza, pp 160–164

Jegede P, Yakubu A, Musa IS, Vincent ST, Shoyombo AJ, Alabi OO, Wheto M, Adebambo AO, Popoola MA (2024) Fertility, hatchability, and prediction of egg weight from egg quality indices of Nigerian indigenous and exotic helmeted guinea fowls. Poultry 4(1):1

Jouffroy D'Abbans I (1996) Dindon et Pintade (Elevage simplifié et rapport). La Maison Rustique, Paris

Korankye O, Kagya-Agyemang JK, Annor SY, Asabere-Ameyaw A, Kyere GC (2020) Effects of season, graded dietary protein and artificial insemination on semen characteristics and reproductive performance of indigenous Guinea fowl (*Numida meleagris*). Bull Anim Health Prod Afr 68(2):251–261

Krunt O, Zita L, Kraus A, Okrouhlá M, Chodová D, Stupka R (2021) Guinea fowl (*Numida meleagris*) eggs and free-range housing: a convenient alternative to laying hens' eggs in terms of food safety? Poult Sci 100(4):101006

Lake PE (1971) The male in reproduction. In: Bell DJ, Freeman BM (eds) Physiology and biochemistry of the domestic fowl. Academic Press, New York

Le Coz DJ (1992) L'élevage de la pintade. Edition point vétérinaire/Maison Alfort. 252p

Leclercq, B. (1982). Alimentation des futurs reproducteurs pintades extrait de fertilité et alimentation des volailles. INRA éd, Versailles, 101–111

MAPA (Ministry of Agriculture, Fisheries and Food/Inprovo) (2004) Guide of good hygiene practices in laying poultry farms. MAPA, Madrid

Muriel A (2005) From Guinea fowl to hatching egg production. MG Livestock World 16(179):60–64

Muriel A, Serrano A (2006) Reproductive and incubation management in Guinea fowl. MG Livestock World 17(190):46–52

Muriel A, Serrano A (2007) Results of an artificial insemination programme in guinea fowl. ITEA (Información Técnica Económica Agraria). Animal Production (Spain) 28(1):66–68

Nwakalor LN, Okeke GC, Njoku DC (1988) Semen characteristics of the Guinea fowl *Numida meleagris meleagris*. Theriogenology 29(3):545–554

Offiong SA, Abed SM (1980) Fertility, hatchability and malformations in Guinea fowl embryos as affected by dietary manganese. Br Poult Sci 21(5):371–375

Pérez FP (1966) Reproducción e inseminación artificial ganadera. Ed. Científico-Médica, Barcelona, pp 521–542

Sauveur B, Reviers M (1992) Reproduction of birds, 2nd edn. Mundi-Prensa, Madrid, pp 193–237

Woernle H (1994) Diseases of poultry. Editorial Acribia, Zaragoza

Feeding

5

Pedro Emilio Rosado Salgado, Francisco Pérez Rojas, and Modesto Senero Fernández

Abstract

The development of guinea fowl requires a certain amount of feed in terms of quantity, content and presentation depending on the age of the bird—chicks, juveniles and adults—as well as sex and productive purpose, either as breeders or for meat. It is recommended that the feed for breeders should contain sufficient energy and protein—18% protein and 2853 kcal of metabolisable energy—to maintain their vitality and optimum egg production, but without generating fat accumulation. Regarding minerals, it is recommended to monitor calcium, phosphorus and manganese intake, as breeding guinea fowl are very sensitive to deficiencies. For guinea fowl intended for meat production, their diet is mainly determined by their age and the farming system chosen: intensive or extensive. The experimental results obtained with these two production systems determine that, in semi-arid areas, intensive farming is recommended in autumn and winter and extensive farming in spring and summer. A very important conclusion derived from the experiments carried out, related to the presence of Ω-3 in the thighs of the carcasses, is the significant correspondence that appears between the content of this fatty acid and the nature of the feed used (organic, commercial, composed of chickpeas or sunflower).

P. E. R. Salgado (✉)
Locust Biological Control Working Group, Department of Plant Pathology, CICYTEX (Scientific and Technological Research Centre of Extremadura), Guadajira, Badajoz, Spain
e-mail: perosados01@educarex.es

F. P. Rojas · M. S. Fernández
CICYTEX (Scientific and Technological Research Centre of Extremadura), Guadajira, Badajoz, Spain

J. del Moral de la Vega, J. del Moral Martínez (eds.), *The Guinea Fowl (Numida meleagris)*, https://doi.org/10.1007/978-3-032-11654-3_5

Fig. 5.1 Batch of guinea fowl in their breeding park

Keywords

Feeding · Feed · Performance · Lighting · Meat carcasses

There are several factors that influence the correct feeding of guinea fowl; these include the age of the animals, the sex and the intended use of the bird (future breeders, breeders and poultry for meat production).

The presentation of the feed is recommended to be in meal up to 2 weeks of age and, from then on, to progressively change to pelleted feed (Blum 1979; Leclercq 1982a; Le Coz-Douin 1992).

The growth of guinea fowl is divided into three periods:

- 1st period: from day 1 to week 4.
- 2nd period: 4th to 8th week (Fig. 5.1).
- 3rd period: from the 8th to the 14th week approximately.

Feeding Management of Future Broodstock

It is important to monitor the weights of future broiler chicken breeders from a very early age. To this end, it is advisable to control the protein supply in the feed (low protein feed), in order to avoid excessive development of the young guinea fowl females. This control of protein supply will allow savings to be made in the future on the birds' maintenance ration during the laying period, thus improving performance. In the case of laying hen strains, feed supplies before and during their productive stage are well studied; however, the opposite is true in the case of guinea fowl.

The feeding recommendations for guinea fowl intended for breeding are as follows:

- Up to 6 weeks of age, a starter feed with 3100 kcal ME/kg and 20% PA (crude protein) is provided.
- From 6 weeks of age, a feed with 2700 kcal ME/kg and 18% PA is used; feed rationing is recommended from the 13th week of age to avoid excessive fattening of the future breeding guinea fowl and thus achieve better laying performance.
- After 23 weeks, the birds shall be fed with feed containing 2850 kcal ME/kg for females and 2700 kcal ME/kg for males.

  During laying, feeding should be ad libitum, but with slight rationing to avoid excessive weights that negatively influence laying.

According to studies carried out by the INRA (Institut National de la Recherche Agronomique) in France (INRA 1975, 1985, 1989), it can be concluded that the guinea fowl, during its growth period, reacts to feed in the same way as the chicken. This type of management allows a better control of the growth rate, significant feed savings and favourable consequences on laying and fertility, a practice which avoids fat deposits in the female from the age of 10 weeks and reduces mortality at the time of laying. The slight delay in sexual maturity can be easily compensated by the lighting programme.

One condition that is advisable to apply during the rearing of breeding stock is the separation of males and females in separate pens, in order to achieve a more homogeneous flock and to avoid competition between males and females in phases when feeding is restrictive.

It can be concluded that, in order to achieve a batch of breeders with optimal characteristics, it is necessary to apply both qualitative feed rationing (restriction of protein intake) and quantitative rationing during the growth of the birds.

Feeding of Guinea Fowl Breeders

The studies carried out by INRA (INRA 1975, 1985, 1989) on this subject have served to determine the energy, protein, amino acid and mineral requirements of breeders.

- Energy concentration of diets

  Breeding guinea fowl decrease their laying when their diet is less than 2850 kcal/kg; moreover, the size of the eggs produced at the onset of laying is slightly smaller.

- Protein requirements
  The optimal coverage of the protein requirements of breeding guinea fowl is around 18% protein and 2853 kcal of metabolisable energy.
- Phosphorus requirements
  There is a specific need for phosphorus to improve egg hatchability (around 0.4% of the ration).
- Calcium requirements
  The guinea fowl egg is characterised by a thicker shell than the chicken's egg; it accounts for approximately 15% of the egg's weight, compared with 9% in chicken's.
  According to various authors, the level of calcium in the diet should be around 4%. Extra calcium in the form of shells does not seem to have a significant effect on the guinea fowl, and it is preferable to provide calcium in the feed mixture.
- Manganese requirements
  It is now known that a manganese deficiency in the diet of guinea fowl (less than 65 mg/kg) causes a drop in the birth rate (Leclercq 1982b; Leclercq and Sauveur 1982).

Feeding of Guinea Fowl for Meat Production

- Energy level of the feed
  For the 1st period, a commercial feed in the form of meal with 3100 kcal ME/kg can be provided; for the 2nd and 3rd period, a level of 3150–3300 kcal ME/kg is recommended.
- Protein level of the feed
  The following percentages are recommended for each period: 25% for the 1st, 22% for the 2nd, and 20% for the 3rd.
  In order to assess whether the poultry feeding programme is correct, the Conversion or Transformation Index (kilograms of feed needed for the animal to gain 1 kg) is used. In Extremadura, the values of the Conversion Index range between 3.5 and 3.8, depending on whether the guinea fowl are reared indoors or outdoors (Fig. 5.2).

- Influence of feed energy content and protein level on growth

Guinea fowl are less sensitive than chickens to variations in energy intake; however, they are quite sensitive to variations in protein intake.

At a given protein level, high and low energy feeds provide a weight gain with little difference. The need for a high calorie intake only concerns the first four weeks of the guinea fowl's life.

Guinea fowl adjust their intake to meet their energy needs for maintenance and growth and are unable to take in excess calories for conversion into fat.

Protein intake from the age of four weeks onwards determines the growth rate. Guinea fowl grow very slowly when the protein supply is insufficient during this period, and as a result, there is a significant delay in growth. To avoid such a delay,

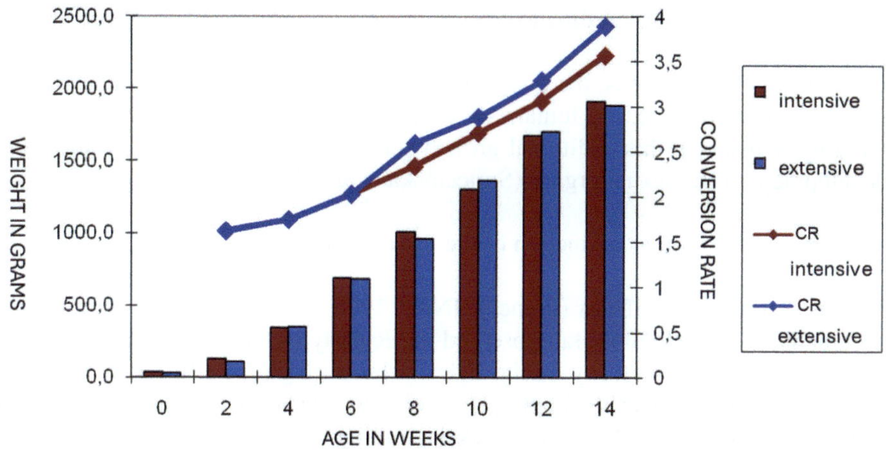

Fig. 5.2 Conversion and growth index of guinea fowl reared using intensive and extensive farming systems, respectively, in Extremadura

Table 5.1 Protein and amino acid requirements of quails according to Leclercq (1982a, b) and Leclercq and Sauveur (1982)

Age in weeks	Proteins	Lysine	Sulphur amino acids
0–4	43.4	2.37	1.68
4–8	50.0	2.51	2.32
8–11	56.3	2.82	2.73

it is necessary to have diets that cover the amino acid requirement during the growth stage of guinea fowl (Alabi et al. 2023; Hounkpêvi et al. 2024a).

Sulphur-containing amino acids are the first limiting factor in the maize-soybean diets commonly used in France (Table 5.1).

- Vitamin requirements

Several studies on the vitamin requirements of guinea fowl have shown that during the early stages of growth, they show similar needs to growing chickens. A high niacin intake is recommended in the early stages to avoid problems with perosis. From 6 weeks of age onwards, the vitamin requirements are significantly reduced.

- Mineral requirements

The mineral requirements of guinea fowl are similar to those of chickens.

- Presentation of the food

The first feed should be presented in crumbs and from the age of 2 weeks onwards in the form of pellets (2.5 mm). However, guinea fowl, unlike chickens, can be fed very well on meal up to the age of eight weeks; neither the growth rate nor the feed conversion rate is affected, as long as the guinea fowl are not wasting feed.

- Body composition according to sex

Up to the age of 12 weeks, the body development of guinea fowl is similar in both sexes. Beyond this age, the female shows greater live weight gains; when sexual maturity is reached, this additional growth corresponds only to fat deposition and development of the sexual organs (Śmiecińska et al. 2022, 2024).

- Body composition as a function of the light programme

Studies carried out by INRA (France) (INRA 1975, 1985, 1989) tried to relate the accumulation of abdominal fat, observed particularly in female guinea fowl at the beginning of sexual maturity, to the light cycles during production periods. They concluded that a decreasing light cycle seems to reduce the amount of fat; other studies have concluded that continuous decreasing lighting reduces the amount of abdominal fat by about 12% in males and 23% in females.

This technique could be used to reduce fattening and improve slaughter performance, among other improvements (Hounkpêvi et al., 2024b).

- Pigmentation of the guinea fowl carcass

The colouring of the guinea fowl skin is an important factor to take into account when marketing this product.

Depending on the area or type of market the product is aimed at, it may have one type of pigmentation or another.

Pigment supplementation provided through feed:

The main materials likely to be used in the guinea fowl feed as a source of pigment are yellow maize, maize gluten, alfalfa meal and alfalfa protein extract.

These are natural products in which the pigment content is variable and depends on the industrial processing system used in the production of feed, the conditions and storage time.

- Recommendations to avoid the production of poultry with fat accumulation

The problem of fattening in guinea fowl from the age of 12 weeks onwards must be taken into account, because if we are only concerned with gaining weight, we can end up producing animals that are too fatty for the consumer. It is possible to modify the production conditions in order to reduce lipid deposition: avoiding too energetic feeds, not rearing females beyond 12 weeks of age (separate the sexes from this age) or using a lighting programme that delays sexual maturity.

- Yield of the guinea fowl carcass according to farming system and time of year

Studies carried out on a total of 200 carcasses of guinea fowl, reared under two farming systems, extensive and intensive in south-western Spain (López Parra et al. 2005; Tejerina et al. 2009), have shown that, under the climatic conditions existing in south-western Spain (semi-arid), the rearing of guinea fowl in the extensive

Fig. 5.3 14-week-old guinea fowl in their enclosure (extensive system)

system (Fig. 5.3) is the most suitable during the warmer months of the year (spring and summer), as it equals or even exceeds the yields of the carcass produced in the intensive system, during the same time period (Fig. 5.4). This may be due, among other factors, to the fact that during this period, especially in spring, guinea fowl have access to a wide range of natural resources (vegetation and microfauna) at their disposal which they use as food, thereby increasing yields and the quality of their meat.

- Influence of feeding on the Ω-3 (omega-3) fatty acid content of guinea fowl meat

The composition of the feed ration determines the content of fatty acids in the carcasses of guinea fowl. In relation to this statement, the Ω-3 fatty acid content of the carcasses of four batches of guinea fowl, each reared on a different type of feed, two commercial feeds (organic and free-range) and two experimental feeds (feed with chickpea and feed with sunflower), have been compared, modifying qualitatively and quantitatively, in each feed, the source of these fatty acids.

The most significant aspect of the analyses carried out on the carcasses of guinea fowl fed with four types of feed, in order to determine the Ω-3 content, is that the birds fed with organic feed had approximately 300% more Ω-3 in the thigh than those fed with the other feeds. These results confirm the studies carried out by numerous authors who state that a change in poultry feed produces a change in the composition of the poultry meat (Fig. 5.5).

With the methodology developed in the study, it is not possible to relate the quality of the carcasses obtained by their Ω-3 content with the nature of the plant material formulated in the feeds tested. Nevertheless, it may be suspected that the higher

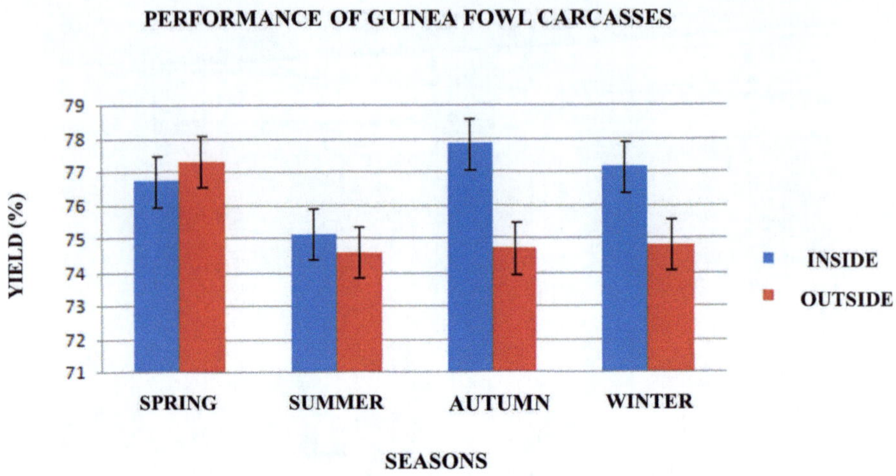

Fig. 5.4 Carcass yield of guinea fowl reared under extensive (outdoor) and intensive (indoor) systems, in four periods of the year

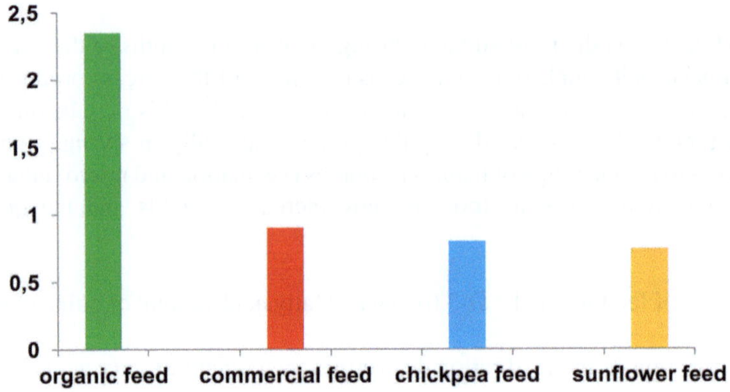

Fig. 5.5 Average thigh Ω-3 values of birds fed different feeds

Ω-3 content of the carcasses is due to the presence of soya, a point that some researchers confirm in relation to chickens, and which will have to be verified with other studies with respect to guinea fowl. On the other hand, the Ω-3 content of the thigh of guinea fowl fed with any of the feeds tested is higher than that established as normal for poultry (0.63 of Ω-3 in the thigh).

The results obtained in south-western Spain coincide with experiences already published by various authors, which demonstrate the close relationship that exists between poultry feed and carcass quality. This encourages further study in this direction, in order to find feeding systems and types of feed with which to produce carcasses of excellent quality (Blum 1979; Castelló 1991; Le Coz-Douin 1992).

Competing Interests The authors have no conflicts of interest to declare that are relevant to the content of this chapter.

References

Alabi OO, Abang C, Olajide OP, Animashahun RA, Etta-Oyong SO (2023) Growth performance and carcass traits of indigenous Nigerian Guinea fowl fed on different dietary protein levels. J Adv Vet Anim Res 10(3):403

Blum JC (1979) Food requirements and body composition of the Guinea fowl. L'élevage de la pintade Session ITAVI

Castelló JA (1991) Chicken meat production. Real Escuela de Avicultura, Barcelona (Spain)

Hounkpêvi JA, Adjei-Mensah B, Adjibodé AG, Tona K, Koutinhouin B, Pitala W (2024a) Dietary protein levels during 12 to 26 wk improve the growth performance, bone quality, and testosterone in Pearl Gray male Guinea fowl (*Numida meleagris*). Poult Sci 103(1):103173

Hounkpêvi JA, Adjei-Mensah B, Parobali T, Tona K, Koutinhouin B, Pitala W (2024b) Photoperiod manipulation corrects the negative effect of low dietary crude protein levels during weeks 12 to 26 on bone quality and fertility of male Guinea fowl (*Numida meleagris*). Poult Sci 103(7):103743

INRA (1975) Institut national de la recherche agronomique. Les volailles de consommation. Séance de travail 16 et 17 octobre 1975, INRA éd. Station de recherches avicoles, 37380 Monnaie

INRA (1985) Institut national de la recherche agronomique. Feeding of Guinea fowl. INRA éd. Station de recherches avicoles, 37380 Monnaie

INRA (1989) Institut national de la recherche agronomique. L'alimentation des volailles. Le pintadeau de chair. Juin 1989, INRA éd. Station de recherches avicoles, 37380 Monnaie

Le Coz-Douin J (1992) L'élevage de la pintade. Editions du Point Vétérinaire, Maisons-Alfort, France

Leclercq B (1982a) Fertilité et alimentation des volailles. INRA éd. Versailles (REPRODUCTION AND FEEDING)

Leclercq B (1982b) Alimentation des futurs reproducteurs pintades extrait de fertilité et alimentation des volailles. INRA éd, Versailles, pp 101–111

Leclercq B, Sauveur B (1982) L'alimentation des reproducteurs pintades en cours de ponte, exrait de fertilité et alimentation des volailles. INRA éd., Versailles

López Parra MM, García Torres S, Tejerina Barrado D, Blanco Martín T, Moral JD (2005) Progress of results on the quality of Guinea fowl meat in Extremadura. ITEA (Información Técnica Económica Agraria). Animal Production (Spain), (262)

Śmiecińska K, Stępień A, Kubiak D (2022) Effect of variety and sex on the carcass and meat quality traits of Guinea fowl (*Numida meleagris* L.). Animals 12(21):2916

Śmiecińska K, Stępień A, Daszkiewicz T (2024) The fatty acid profile and the quality of breast and leg muscles in female and male pearl gray Guinea fowl (*Numida meleagris*). Poult Sci 103(3):103385

Tejerina D, Lopez-Parra MM, García-Torres S (2009) Potential use of near infrared reflectance spectroscopy to predict meat physico-chemical composition of Guinea fowl (*Numida meleagris*) reared under different production systems. Food Chem 113(4):1290–1296

Health of the Guinea Fowl

6

Pedro Emilio Rosado Salgado

Abstract

Guinea fowl, unlike other farm-raised birds, show few health problems, although they are not immune to them, most of which are described in this section, which describes the symptoms shown by affected birds, their specific aetiologies, the prophylaxis that prevent them and the therapeutics aimed at curing them. This inventory is preceded by a description of some aspects of the anatomy and physiology of the guinea fowl considered important because of their relationship with the different pathologies. Numerous species of microorganisms (viruses, bacteria, mycoplasmas and pathogenic nematodes) and arthropods (mites and insects) are mentioned as disease-causing agents. Moreover, there is a description of other diseases derived from inadequate feeding, poor management, relations between animals (pecking) and dislocations. Body examinations or blood and faecal analyses necessary for the identification of the respective aetiologies need to be done by various methods which are described in this section. Finally, the mandatory biosecurity measures on farms, as well as the necessary vaccinations and how they are carried out, are detailed in this chapter. In order to be able to recognise the described diseases at first sight, tables have been prepared which show the symptoms of the guinea fowl affected by each disease in order.

Keywords

Physiology · Pathology · Pathogenic agents · Prophylaxis · Treatment

P. E. R. Salgado (✉)
Locust Biological Control Working Group, Department of Plant Pathology, CICYTEX (Scientific and Technological Research Centre of Extremadura), Guadajira, Badajoz, Spain
e-mail: perosados01@educarex.es

J. del Moral de la Vega, J. del Moral Martínez (eds.), *The Guinea Fowl (Numida meleagris)*, https://doi.org/10.1007/978-3-032-11654-3_6

In any livestock farm, the knowledge and control of infectious processes that may occur are of vital importance to avoid serious problems and to ensure quality and safety of production from a sanitary point of view.

A poultry farm has a number of points in common with other livestock farms, but at the same time has some very unique features.

Birds, anatomically and biologically, are different from mammals; therefore, we will have to adapt the management guidelines to their characteristics, in order to avoid health problems and to ensure their maximum welfare.

The guinea fowl is a bird that adapts well to any management system, whether extensive or intensive, resisting many infectious processes that affect other species more virulently, for example, laying hens or chickens for meat production. This does not mean that the sanitary guidelines in a guinea fowl farm should be relaxed or that they should be the same as in any other type of poultry species farm.

In this chapter, we will give a brief description of the anatomy of the guinea fowl and address the particularities of management that can influence the health status of these birds, including a brief description of the aetiological agents that cause disease in the species.

Some Important Aspects of Guinea Fowl Anatomy and Physiology

Skeleton

Primarily, the skeleton of birds is adapted for flight. It is early ossified and has a series of well-developed structures for this purpose, such as the shoulder girdle and the vertebral column; in addition, the bones of the head and the humerus have pneumatic chambers to lighten the bird's weight.

The guinea fowl performs short, powerful flights, similar to those of other similar birds, such as the partridge or pheasant.

In general, the vertebral column of birds provides great mobility to the head due to the 13–18 vertebrae that support it, depending on the species. In the case of the guinea fowl, the head has a horny, helmet-shaped protuberance that adorns the top of the head, which in the case of the male is slightly more developed.

The bones of the legs are strong and well adapted for walking and running, with three toes forward (second, third and fourth) and the first toe pointing backwards.

The legs of cockerels are longer and stronger than those of the female guinea fowl.

Respiratory System

In addition to the small, unlobed lungs attached to the ribs and vertebrae in the thoracic cavity, birds generally have air sacs, which are evaginations of the bronchi located in the body cavities and long bones. Another anatomical peculiarity of birds is that they do not have a diaphragm.

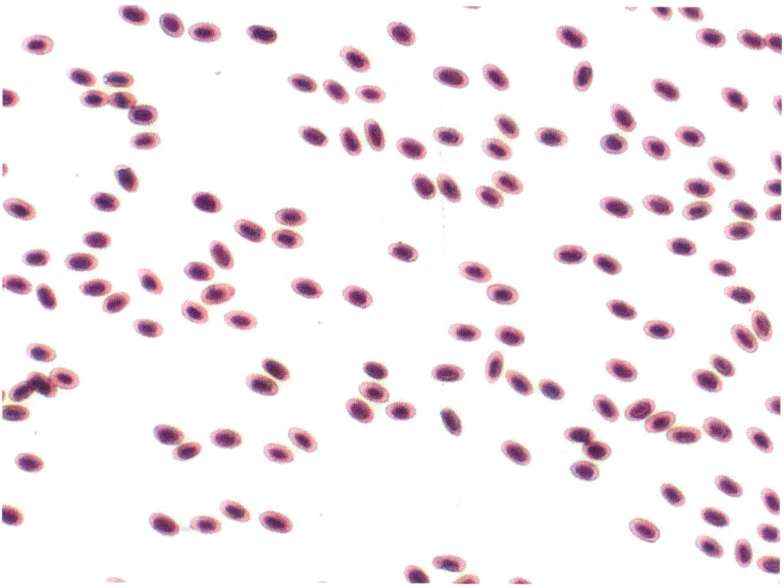

Fig. 6.1 Red blood cells in guinea fowl seen under a microscope. Note the presence of the nucleus, characteristic of all birds

The respiratory system in birds is very efficient, so they can easily obtain oxygen for activities such as flight.

A healthy lung is pinkish in colour, has a spongy appearance and is devoid of spots, secretions, etc., presenting a uniform appearance.

The avian heart has four separate chambers (two atria and two ventricles), and it is the right aortic arch that gives rise to the systemic circulation (unlike in mammals, where the left aortic arch is involved). The red blood cells of birds have a nucleus and are oval (Fig. 6.1).

Digestive System

Birds select their food on the basis of shape and colour, preferring, in the case of domestic species, feed in granulated form to commercial meal.

It is interesting to note that in many birds, except pigeons and waterfowl, the sense of taste is not highly developed, so bitter-tasting medications can be administered via drinking water.

The beak is the organ with which birds pick up food; in the case of guinea fowl, it is robust, with shades varying from dark brown to ivory.

Food reaches the crop through the oesophagus. The crop (Fig. 6.2) is a dilation of the oesophagus where food is softened before passing to the glandular stomach or proventriculus (initiates digestion with its secretions) and the muscular stomach or gizzard, where digestion is completed and food is crushed with the help of

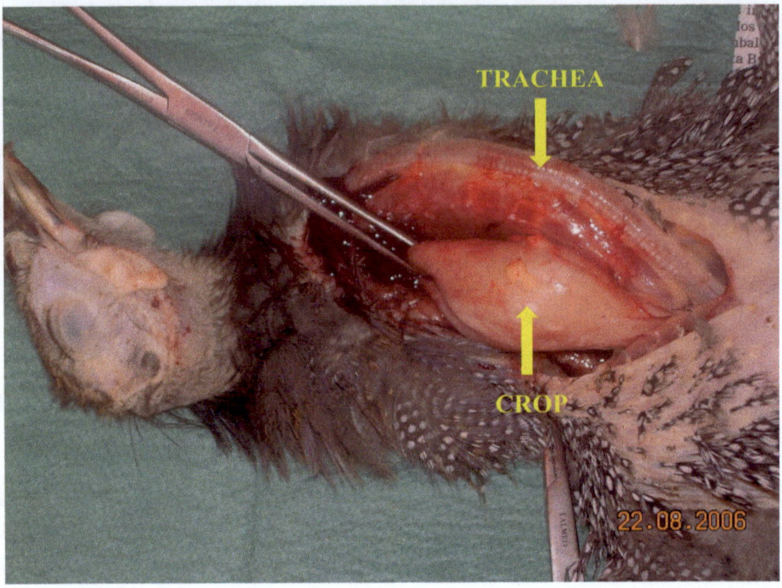

Fig. 6.2 Location of the crop and trachea in guinea fowl

pebbles ingested by the bird. These pebbles are necessary for the proper functioning of the gizzard, thus avoiding digestive disorders. In the case of guinea fowl accessing outdoor parks, they will obtain these pebbles from the ground of the parks themselves.

As for the viscera attached to the intestinal tract, the liver (Fig. 6.3) has two main lobes, right and left. The hepatopancreatic duct, which drains bile directly into the duodenum, is connected to the left lobe. Two short hepatocystic ducts, which carry bile to the gall bladder (Fig. 6.4), are connected to the right lobe. This feature is absent in certain psittacine species, pigeons and ostriches. From here, the bile will be discharged into the duodenum via the cysticenteric duct.

The pancreas (Fig. 6.5), which is enveloped by the small intestine, specifically by the duodenal loop, appears as an elongated cream-coloured gland.

The small intestine (Fig. 6.6) extends close to the cloaca. As a result of this, in cases of "pecking", a problem that can occur in guinea fowl housed in laying batteries, neighbouring birds can easily externalise the intestinal packet of their victims.

Guinea fowl have two caeca arranged in parallel; this is where the bacterial flora that performs complex digestion and vitamin production is found.

At the end of the colon, we find the cloaca, where the ureters and the oviduct or sperm ducts, in the case of the male, end.

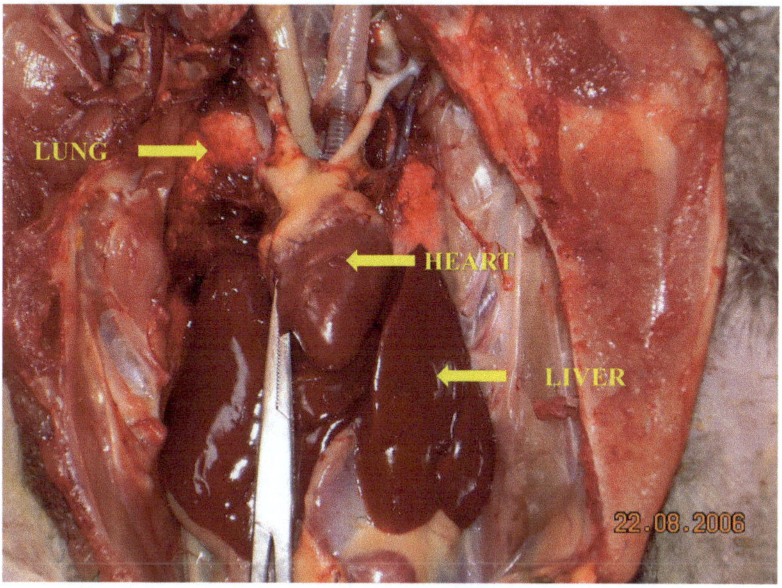

Fig. 6.3 Anatomical location of the heart, liver and lungs in guinea fowl

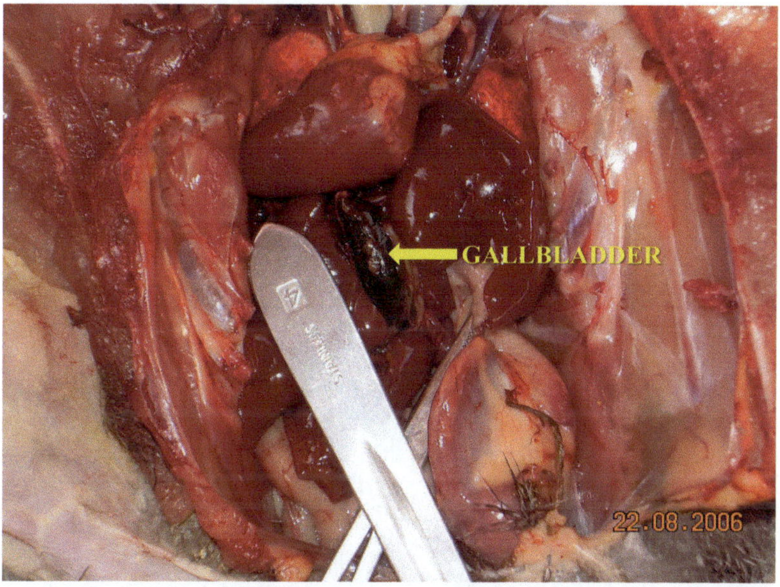

Fig. 6.4 Gallbladder

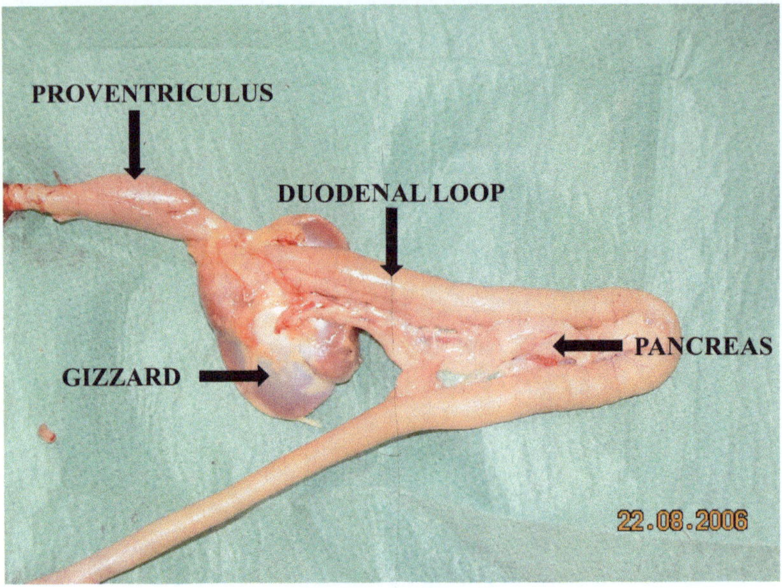

Fig. 6.5 Location of the glandular stomach (proventriculus), muscular stomach (gizzard), pancreas and duodenal loop of the small intestine surrounding the pancreas

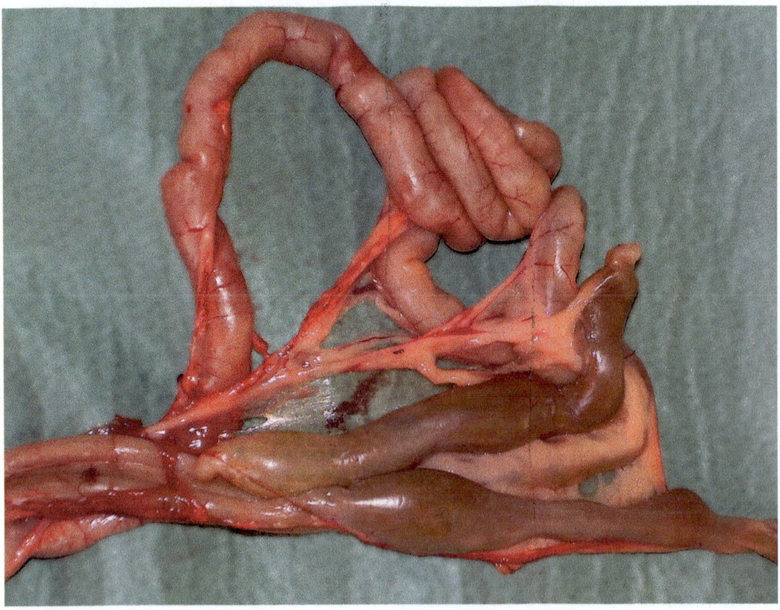

Fig. 6.6 Structure of the small intestine and caeca of the guinea fowl

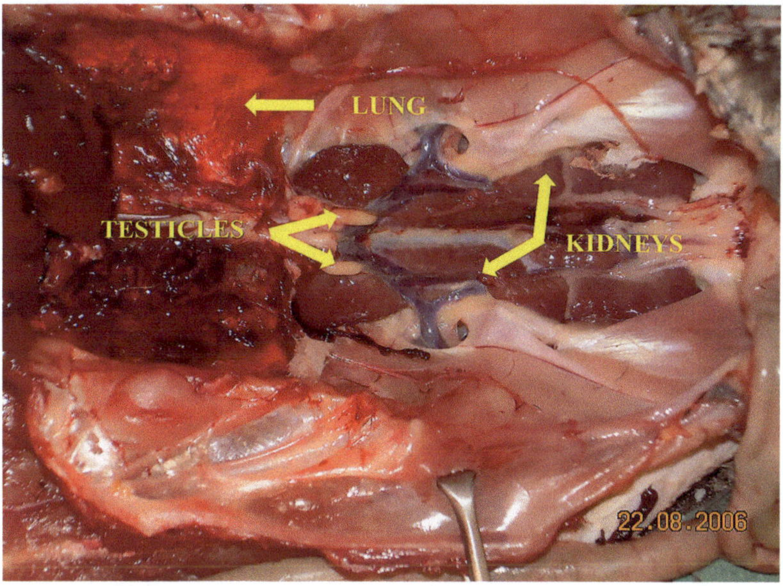

Fig. 6.7 Lungs, testicles and kidneys in a guinea fowl cock

Excretion and Reproduction

The two kidneys (Fig. 6.7), which are situated on the ventral side of the synsacrum and ilium bones. Are elongated and divided into three lobes; the ureters empty into the cloaca (urodeum). Birds do not have a urinary bladder.

The testes (Fig. 6.7) are housed in the abdominal cavity. There are no accessory sex glands. The male copulatory organ is usually rudimentary. The penile papilla in other species can reach greater lengths. Guinea fowl possess a penis that can reach about 3.5 cm in length.

In hens, only the left ovary and oviduct are developed and have a cluster-like appearance. The oviduct consists of the following parts: infundibulum, magnum, isthmus, ampulla, uterus and vagina. The latter is connected to the urodeum on the left side. It is situated in the abdominal cavity, below and in front of the kidneys, to the left of the spinal column. It is situated in front of the lungs, diaphragm, liver and foregut; its dimensions vary considerably, depending on the functional state in which it is found. In the resting state, it appears as a whitish-grey mass about the size of a blackberry, and in the active state, it appears as a bunch of grapes of different sizes, depending on the degree of maturity; it is yellowish in colour with pinkish or dark greyish tinges. The organ is round to polygonal in shape, divided into lobes, with a friable consistency (Fig. 6.8).

The oviduct and uterus are represented by a hollow, tubular, muscular organ, which runs from the ovary to the cloaca, occupying a large part of the abdominal cavity, and is extraordinarily dilatable. It is subdivided into the infundibulum, the

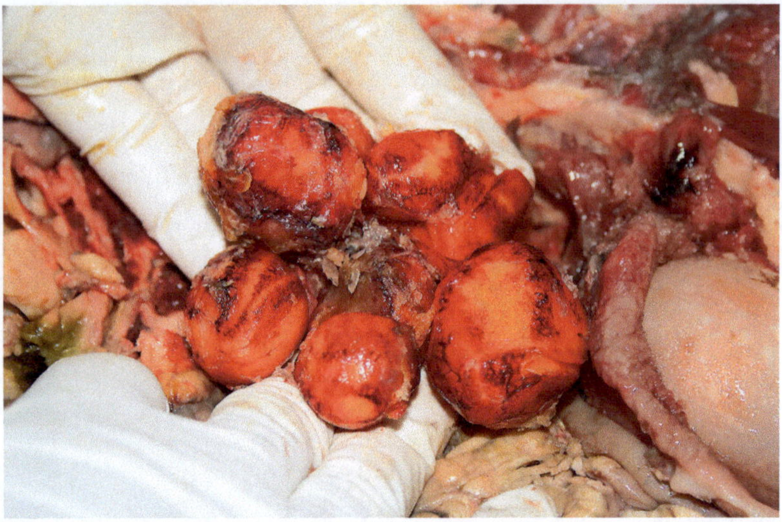

Fig. 6.8 Appearance of ovarian follicles in a female guinea fowl

magnum or albuminiferous chamber, the isthmus, the uterus or calciferous chamber and the vagina.

The vagina is the last part of the oviduct and appears as a short S-shaped tube, the site of the junction between the vagina and the uterus, but its vaginal glands are the primary storage sites for spermatozoa from natural or artificial fertilisation.

It has been sufficiently demonstrated that birds, unlike mammals, lack accessory sex glands, such as Cowper's glands, seminal vesicles, prostate and urethral glands.

The genital organs of the male consist of the germinal glands, represented by the testes, the epididymis, the semen excretory apparatus and copulatory organ and the cloaca (Sturkie 1968; Hoffman and Völker 1969; Sisson and Grossman 1982).

Sanitary Management of Guinea Fowl

In guinea fowl farms, there are some operations that must be considered almost indispensable if a good health status of the birds is to be achieved.

Group Management of Birds

To capture guinea fowl in groups, the use of hinged fences to trap the animals in the corners of the breeding and rearing pens is recommended in both outdoor and indoor parks, in order to immobilise as many birds as possible. Capture should be rapid to avoid crowding and stress-related deaths. Also, excessive numbers of birds should not be rounded up at one time, as this will encourage the aforementioned accidents.

Fig. 6.9 Breeding pen for guinea fowl with rounded corners to prevent suffocation or crushing

In the case of indoor breeding farms, it is advisable to reduce the light intensity in order to reduce stress on the animals and facilitate their capture.

In indoor breeding pens, it is advisable to soften the corners (Fig. 6.9) to prevent suffocation due to crowding. For this purpose, the corners should be rounded in such a way that the guinea fowl can escape in the event of crowding.

Another very important measure is to secure the feeders and drinkers to the roof to prevent birds from tipping them over; for this purpose, it is advisable to include in the construction of the rearing pens a metal beam running through the pen from wall to wall, from which the feeders and drinkers can be hung (Fig. 6.10).

Individual Handling of the Guinea Fowl

In order to carry out any sanitary operation on them (extraction of faecal samples, blood, vaccinations, body examinations, etc.), and due to the nervous nature of guinea fowl, it is necessary to be careful when holding them. To this end, the bird should not be held only by its legs (Fig. 6.11), and it is advisable to hold the guinea fowl by its wings as well (Fig. 6.12). To carry out this type of operation, it is advisable to have a table nearby on which to place the bird and the auxiliary material.

Fig. 6.10 In indoor breeding pens, it is advisable to install a beam from which to hang feeders and drinkers

Body Examinations

It is not difficult to notice that a sick bird does not move like a healthy bird; it tends to be drowsy, keep its eyes closed, show ruffled feathers and breathe with its beak open and even sneeze conspicuously; it also does not feed regularly and even stops eating and drinking.

Once in our possession, signs of disease can be detected if the bird is not alert, if the body condition is not adequate, presenting a too thin appearance and abnormally decreased weight. The nostrils and cloaca area should also be observed; in the first case, if we detect traces of mucus or secretions, we can suspect a respiratory process, and in the second case, if we observe the feathers around the cloaca stained with faeces, we can suspect an enteric process.

Fig. 6.11 Incorrect way to immobilise a guinea fowl

Blood Collection

Blood collection in birds is carried out by various methods, but the most hygienic, least traumatic for the bird and most comfortable for the veterinary professional is the syringe puncture in the brachial vein (Fig. 6.13), on the inside of the wing. To do this, it is advisable to clear the area of feathers and disinfect. The bird should be properly immobilised on a table by an assistant or carer.

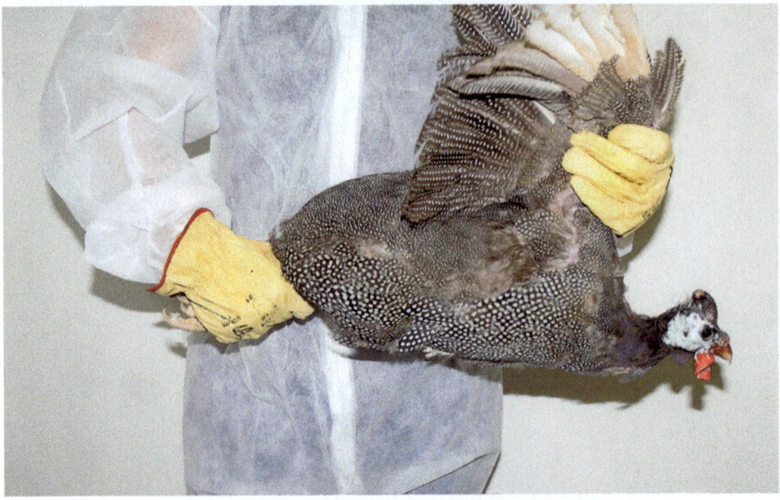

Fig. 6.12 Recommended way to catch a guinea fowl

Fig. 6.13 Blood should be taken from the brachial vein (wing vein) of guinea fowl

Faeces Collection

Collection of faeces can be done collectively, in the faeces collection pits, or fresh samples can be taken from groups of animals in the batteries or from the boot swabs of the farm operators. If the faeces sample is taken individually, it should be taken by means of a cloacal swab, which should be placed in an airtight and sterile container (Le Coz-Douin 1992; MAPA 2004, 2005).

Pathology

General Signs of Disease

When a bird presents some kind of pathology, some characteristic signs can be observed that can help us detect the existence of a health problem.

Among the most common signs, we can observe early on that birds do not consume the usual amounts of feed and water; in the case of water, consumption may be below normal or above normal.

Among other signs of illness, we will observe that a sick bird is almost immobile, standing apart from its companions, cornered, with its plumage ruffled and its neck tucked up, its eyes often closed and in a state of drowsiness.

There are many signs of respiratory disorders, such as eye discharges, conjunctivitis, nasal secretions, difficulty and noisy breathing.

Manifestations of enteric processes include liquid, fetid and abnormally coloured faeces, and the feathers around the cloaca are stained with the remains of defecation or even blood.

Another group of manifestations that can help identify a health problem on the farm are found on the skin on the ridge or skin of the head; due to the fact that the body of the birds is covered with feathers, in these areas in many species, including the guinea fowl, we can observe discolorations or abnormal colours that reveal disease.

Main Disease-Causing Agents in Poultry

Poultry on a farm are exposed to a variety of biological and non-biological agents that cause disease. Controlling these factors will be essential to optimise the profitability of the farm, because if the pathological process spreads throughout the farm, it will considerably increase production costs and may even compromise the viability of the farm and put the health of consumers at risk.

Among the causes of disease in productive poultry, we can distinguish between external and internal causes.

External or acquired causes can be divided into those caused by biological aetiological agents, such as bacteria, viruses, fungi, parasites and mycoplasmas, as the most prominent and those caused by deficiencies in handling and nutrition.

Management problems in breeding and rearing are almost exclusively limited to the first 6 weeks of life, which is the most critical period in production. After this time, normally, if the guidelines applied are correct, and with a minimum level of common sense, there are usually no serious problems, due to the tremendous hardiness and adaptability of this bird to our agro-ecosystem.

In the case of females housed in batteries, attention must be paid to feeding, drinking water and medications, taking into account that good performance of the birds thus confined will be dependent on their correct administration. If these measures are carried out incorrectly or deficiently, they will cause serious problems for the farm and will affect not only a single animal but all of its neighbours in the battery, since they share the same facilities and the same management guidelines. For this reason, great care must be taken with the aforementioned aspects and with the environmental conditions of the breeder house (humidity and temperature), with the personnel who have access to these facilities, with the preventive measures, etc. It is important to remember that this area (breeder house) is the heart of the farm and that the success of the rest of the production process of a guinea fowl farm depends on its proper functioning.

Table 6.1 summarises some of the symptoms that can alert us to poor management or feed deficiencies in poultry on a farm.

The internal causes of diseases derive from genetic defects that can occur in certain animals. This type of problem is fairly well controlled in common industrial poultry farms, and does not usually cause serious disorders, but in our area, as the industrial production of guinea fowl is not so developed, it is important to control the offspring in order to avoid possible alterations that make the birds unsuitable for production. In the case of guinea fowl, problems of congenital origin are not usually frequent if the selection of breeders is carried out with the precaution of eliminating those birds that have the capacity to transmit these types of anomalies to their offspring (Le Coz-Douin 1992; Woernle 1994; McMullin 2020).

Table 6.1 Frequent symptoms that may be present in laying hens, indicative of poor management or nutritional deficiencies, according to Prof. Hellmut Woernle

Symptomatology	Likely cause
Dyspnoea (breathing through the beak) and drop in the laying of eggs	High temperature in the poultry house and water shortage
Respiratory diseases	High amount of ammonia in the air in the poultry house
Coccidiosis or mites	Excessive humidity and high temperature in the poultry house
Feather pecking or cannibalism	Excessively hot and humid park climate, excessive lighting
Swollen and infected eyelids	Excess weight in layers, very wet ground and lack of perches in walkways
Rickets in Guinea fowl	Vitamin D and mineral salt deficiency
Intestinal disorders	Defects in rationing, germs in the feed
Fatty liver syndrome	Unsaturated fatty acid deficiency, choline and protein deficiency
Digestive disorders, stomach obstruction	Lack of stones in the muscle stomach
Signs of decalcification of bones, paralysis of layers	Mineral salt deficiency in laying Guinea fowl

General Characteristics of Pathogens and the Symptoms They Produce

Bacteria

Bacteria produce their pathogenic action by colonising and causing the destruction of the tissues where they are found. The whole process is accompanied by inflammation, oedema, bleeding and abnormal colouring of the organs where they exert their pernicious action, and in the case of generalised dissemination throughout the organism, they produce what is known as septicaemia, causing immediate death with generalised signs.

The main weapon against bacteria is hygiene and disinfection, in addition to establishing strict management protocols on the farm to prevent the spread of the disease. Once the disease is established in a flock, therapy can be carried out by administering antibiotics, mainly as chemotherapy products.

Viruses

These microorganisms exert their pathogenic action by colonising tissue cells, and once they reproduce inside the cell, they lyse it, destroying it completely. Affected animals die, but others may remain as asymptomatic carriers, releasing the virus in their exudates or faeces, which can wreak havoc in other groups of birds. Another characteristic of viral infections is that they are often accompanied by secondary bacterial infections, which aggravate the pathological picture.

Parasites

They are divided, for study purposes, into external and internal. The external ones are usually insects or mites that generally affect the plumage and skin tissue. The internal ones include multicellular (nematodes) or unicellular (Protozoa) organisms, which affect various parts of the bird's body, especially the respiratory and digestive systems.

Parasite cycles are usually complex, with phases inside and outside the body of their definitive hosts. To break this chain, it is first necessary to make an accurate diagnosis of the type of parasite that is causing the problem and then to apply the necessary sanitary and hygienic measures, although it is often possible to avoid reaching this point by applying disease prevention protocols.

Deworming products are usually given to the animals in their drinking water and, in the case of internal parasites, in their feed. In the case of external parasites, dewormers are usually applied as part of the disinfection and dis-insectisation measures or when the health vacuum is applied.

Fungus

Among the genera that are diagnosed as animal pathogens, we find, among others, *Candida* and *Aspergillus*, the latter being the one that causes the greatest havoc in birds, especially in the respiratory system and air sacs. Its pathogenic action is carried out directly in the affected tissues or through the production of toxins in the birds feed, producing very serious disorders. To prevent their action, good control over the presence of these types of toxins in feed mills, good storage conditions of the feed, adequate hygiene measures in the animal houses, avoiding excess humidity and cleaning and disinfecting tools frequently are necessary (Cruciani et al. 2022; Ugochukwu et al. 2022).

Mycoplasma

These microorganisms, capable of causing disease in poultry, are characterised by the fact that they have no cell wall and wreak havoc in the respiratory tract of birds. They are difficult to treat, and their persistence is high. Control is based on rigid disinfection measures of the facilities (Jacob and Pescatore 2022; Porter Jr and Morishita 2022b).

When birds show signs of disease, veterinary intervention is necessary. Through the interpretation of these signs, the lesions revealed in the necropsies and the referral of samples to veterinary diagnostic laboratories, the most appropriate treatment and preventive measures against the germ causing the problem can be established. Treatment is indicated in those pathologies where slaughter is not compulsory due to current legislation or for economic or practical reasons. In the case of certain viral infections, treatment is limited to the application of antibiotics to control secondary bacterial infections while waiting for the bird's own organism to overcome the viral infection.

When sending samples to the laboratory, it is necessary and very useful to send a supporting detailed veterinary report.

Table 6.2 highlights some of the identifying alterations of diseases in birds that can be seen, at first glance, in an inspection of sick or necropsied birds.

Pathologies of Viral and Bacterial Aetiology

Avian Encephalomyelitis

Also known as "epidemic scrapie" or "New England disease", a disease caused by a globally distributed, non-enveloped RNA virus of the family *Picornaviridae* and genus *Hepatovirus* (Coles 2009).

Strains are divided into enterotrophic and neurotrophic strains according to their virulence, the most common hosts being chickens, turkeys, pheasants and quails.

Table 6.2 Easily identifiable symptoms of sick birds, attributable to different pathogens, according to Prof. Hellmut Woernle

State of meat	Good in acute diseases of rapid course. Thinning in chronic diseases of slow course. Recognition is by palpation of the chest musculature. The sternum protrudes in thin birds
Body posture	Twisted head in Newcastle disease. Paralysis of the limbs in Marek's disease. Decay in chronic diseases, such as parasitic infestations, leukosis or tuberculosis
Plumage	Observe the presence of external parasites. Presence of bare patches in the "pecking" and cannibalism. Observe the appearance of possible exaggerated moulting
Pupils	Disfigured pupils in Marek's disease
Breathing through the beak	It is produced in inflammation of the respiratory tract, in hot climates to give off heat and is necessary to balance body temperature
Nasal openings	Moist nasal openings in diseases of the respiratory tract, by squeezing the nasal apertures mucus appears, which, when the disease is old, smells bad
Infraorbital sinuses	Lumpy in poultry colds and Turkey sinusitis and Guinea fowl
Pharyngeal cavity	Fibrin deposits in smallpox-diphtheria. Presence of small fibrin grains and defects in pharyngeal mucous membranes in malignant Newcastle disease. Cornification of the excretory ducts of the mucous glands when vitamin A deficiency occurs
Skin colour of the head and chins	Violet in acute and febrile diseases or circulatory disorders, as well as in cardiac and renal diseases; pale in long-term chronic diseases, such as parasitic infestations, tuberculosis and leukosis; presence of caseous deposits in smallpox
Sewer area	In enteric catarrh, the feathers of the cloaca are dirty, with presence of faecal particles, or are agglutinated with inflammatory products, in inflammation of the oviduct. Attention should be paid to the characteristics of the faeces. In severe coccidial infestation, faeces are mixed with blood
Lower extremities	Paralysis in Marek's disease. In perosis in young birds, deviation of the tarsal joint due to slipping of the Achilles tendon. Scabbing of the skin of the limbs in mite infestation of old birds, causing swelling of the fleeces after wound formation and penetration of bacteria
Pubis	Narrowing at cessation of laying
General drowsiness: sadness, decreased feed intake, ruffled feathers	In heavy animals, it may be a sign of the onset of acute bacterial or viral disease or coccidiosis In light animals it can originate in the final phase of a chronic disease (tuberculosis, leukosis, parasitic infestations)
Movement disorders	Paralysis in Marek's disease. It may be a sign of osteomalacia. If head deviation occurs, it may be a sign of slow onset Newcastle disease Extreme flexion of the legs and support on the tarsal joint in disorders such as perosis (displacement of the Achilles tendon from the tarsal joint)

(continued)

Table 6.2 (continued)

Fur and plumage	In sudden cases of death, birds with pale heads may appear. Eggs of certain parasites may appear in the cloacal area: bird mites and lice eggs Fleas may appear in nest boxes Scabies on the dorsal area: cutaneous scabies mites A disease that produces the appearance of "chalky legs" may be caused by mites The appearance of deteriorated plumage and haemorrhagic areas on the back and cloaca may be a consequence of the dryness of the climate in the poultry house Cannibalism, intestine on the outside, through the cloaca, as a consequence of restlessness during laying
Muscles	Pale ochre colour of the chest musculature shows metabolic disorders Muscle haemorrhages in haemorrhagic syndrome and Gumboro disease Pale muscles as a result of blood loss due to red bird mites, internal haemorrhage in tuberculosis or fatty degeneration of the liver Dark-bluish-red colour of the musculature after death with high fever process
Head	Dark-violet-blue colour of the skin of the head in acute infectious disease or poisoning and enzootic cardiac death Pale skin of the head in chronic disease, leukosis and tuberculosis Scabs on the head and horny cap may mean smallpox
Eyes	Clouding of the cornea when the environment in the poultry house is rich in ammonia Pupil deformation in Marek's disease Eyelids sticking together in cold diseases and vitamin A deficiency
Nasal openings	Flux: poultry catarrh, mycoplasmosis, various viral infections of poultry
Infraorbital sinuses	Increased volume in mycoplasmosis and sinusitis
Pharynx, upper and lower larynx	Beak breathing, often accompanied by vocalisation and rales in mycoplasmosis, viral infections and chronic respiratory organ disease Presence of inflamed diphtheroid foci in Newcastle disease Fibrin deposits in smallpox-diphtheria Cornifications of the glandular canals when vitamin A deficiency occurs
Trachea and lungs	Inflammation in mycoplasmosis and viral infections
Lungs	Inflammation with suppuration in *E. coli* infection in the CRD Millet-like granules in lungs of Guinea fowl in *S. pullorum* infection (white diarrhoea in Guinea fowl)
Aerial bags	Caseous (cheesy) fibrin deposits: chronic respiratory tract disease (CRD)
Buche	Paralysis in Marek's disease. Gas formation due to altered feed Clogging of the crop by food retention due to lack of dietary pebbles
Glandular stomach	Inflammation and haemorrhages in haemorrhagic inflammation of glandular ducts in the course of Newcastle disease, intoxications, renal urate and haemorrhagic syndrome; erosions in the passage to the muscular stomach in stress syndrome. Bulging with increased volume in watery Marek's disease, with lymphocyte proliferation

(continued)

Table 6.2 (continued)

Muscular stomach	Relaxed musculature and obstruction, muscle involution with gastric dilatation: sand deficiency (especially in floor and park breeding after ingestion of litter or long fibre vegetables), common in Guinea fowl Erosions with lack of substance in the keratin layer in stress syndrome and eating disorders
Small intestine (outer layer)	Presence of nodules consisting of tuberculosis Juicy tumour formations in myelosis (proliferation of white blood cells) Cumulus streaks in the duodenum due to coccidiosis caused by *Eimeria acervulina* Greyish-white dots in the duodenum: coccidiosis due to *E. mivati* Haemorrhagic spots in the entire duodenum: *E. necatrix* coccidiosis Oedematous and reddened intestine: Coccidiosis due to *E. maxima* Vessels very full of blood in enteritis and circulatory paralysis Abundant hard nodules throughout the intestine and serous membranes: malignant tumours, often originating from the pancreas
Small intestine (internal mucosa)	Haemorrhagic, viscous or fluid intestinal contents: various coccidiosis, enteritis due to intestinal flora disorder and salmonellosis Mucous content in anterior intestine and bile retention: infestation with small, barely visible flatworms, when birds are kept in flocks (infestation with filamentous worms) Presence of round worms, up to 10 cm in length: ascaridiosis Sandpapery appearance of the mucosa, especially of the final intestinal tract: necrotising enteritis (enteritis due to abnormal intestinal flora, abundance of *Clostridium perfringens*) Nail-sized diphtheroid inflammatory foci: Acute Newcastle disease
Cecum	Red bloody content: Caecal coccidiosis due to *E. tenella* Chronic inflammation with caseous fibrin deposits: healing caecal coccidiosis. Contagious inflammation of liver and cecum in the so-called "black head" disease
Straight	Malodorous fibrinous caseous deposits: coccidiosis due to *E. brunetii*
Sewer	Blood-infiltrated lesions, duodenum infiltrated with blood apparent on the outside: cannibalism Clumpy and dirty feathers in the cloacal area: sustained intestinal inflammation and inflammation of the oviduct
Fabricio bag	Increased volume and sometimes fibrinous or haemorrhagic content in pinheads: infectious disease of the bursa (Gumboro)
Ovary	Plethoric blood: acute infectious disease and poisoning with circulatory paralysis Regressing yellow follicles: acute viral infection Degenerated follicles: bacterial infections (*Salmonella pullorum*) Follicle degeneration and tumour formation: ovarian tumours, Marek's disease Accumulations of fibrin, pus and shell masses in the oviduct: inflammation of the oviduct, formation of layers in the oviduct and oviduct-peritoneum inflammation Cloacal cysts, with watery or fibrinous contents: rudimentary right oviduct with cystic degeneration

(continued)

Table 6.2 (continued)

Pericardium	Caseous-fibrinous inclusions: *E. coli* infection in CRD, *S. pullorum* infection in old hens and Pasteurella infection Chalky inclusions of uric acid salts: visceral lithiasis Haemorrhagic spots and discoloured haemorrhages: acute viral infection (Newcastle disease) and bacteriological conditions (*E. coli* septicaemia, fowl cholera)
Myocardium	Whitish nodular cellular inclusions: *S. pullorum* infection in Guinea fowl More diffuse inclusions: Marek's disease Musculature of ochre tone due to fat deposition: myocardial degeneration Spherical heart formation and ochre-coloured myocardium: enzootic cardiac death
Blood vessels	Marked blood repletion, especially in liver and ovary: circulatory paralysis
Spleen	Dry-caseous nodules: tuberculosis Diffuse, whitish, moist lumps: leukosis, and Marek's disease
Liver	Gypsum coating: visceral lithiasis (uric acid and urate deposits) Whitish lardy inclusions or diffuse enlargement of the liver, pale grey: leukosis or Marek's disease Dry caseous nodules, sometimes with ruptured liver and internal haemorrhage: tuberculosis Circular dead areas with congestive halo in the liver: contagious inflammation of the liver—caecum, also known as "black head" (histomoniasis) Light ochre and friable liver, often with haemorrhagic areas or liver rupture and internal haemorrhage: fatty liver syndrome
Gall bladder	Very full: in inflammations of the duodenum, sometimes with capillary worm infestation
Kidneys and ureters	Renal canaliculi and ureters full of uric acid and white urate (healthy ureter is thin and inappreciable): visceral lithiasis, Gumboro disease or water deficiency
Bones and joints	Ribs and long bones arched in keets: rickets Bowed ribs in laying Guinea fowl and protrusion of the rib joint: osteomalacia Manifestations of paralysis: haemorrhage in vertebral canal after fracture between the fourth and fifth dorsal vertebrae Slipped Achilles tendon of the tarsal joint and abnormal posture of the leg bones: perosis Limb bone hypertrophy: osteopetrosis (bone leukosis) Consistent, yellowish-caseous nodules in bone marrow and joint inflammation: various joint infections

Transmission can be either horizontal (faecal-oral route) or vertical (increased embryonic mortality or hatching of infected chicks).

Incubation period is 1 to 2 weeks.

Fig. 6.14 Paralysis and prostration in a group of guinea fowl

Fig. 6.15 Paralysis associated with enteric symptoms (diarrhoea) in a guinea fowl a few days old

- Clinical signs: (chicks 1–3 weeks old)

  1. Ataxia, progressive paralysis, prostration, tremors (Figs. 6.14 and 6.15).
  2. Morbidity up to 50–60%.
  3. High mortality (up to 60%); guinea fowl die from crushing or starvation.
  4. Differential diagnosis with Marek's disease, encephalomalacia (vitamin E deficiency), infectious bronchitis and Newcastle disease.

- Clinical signs: (adult laying hens)

  1. Decreased laying (low mortality)
  2. Increased embryonic mortality (decrease of hatched eggs)

- Prophylaxis
  1. Vaccination of hens before laying (8–16 weeks of age).
  2. Attenuated or inactivated vaccines (laying birds).
  3. In case of an outbreak, cull all affected birds, clean and disinfect with 2% formalin.

Turkey Rhinotracheitis (TRT) and Swollen Head Syndrome (SHS)

Turkey rhinotracheitis and swollen head syndrome in hens are caused by the avian *Metapneumovirus*, an RNA virus belonging to the family *Paramyxoviridae* (Suarez et al. 2020). This disease affects mainly turkeys and hens but can also occur in guinea fowl and guinea fowl farms. The most affected are flocks of growing birds between 8 and 16 weeks of age; at these ages, the morbidity is usually high, reaching even 100%; mortality can be up to 50%, depending on how the flock is managed during the process.

The disease affects the upper respiratory tract, producing aerosaculitis and polyserositis; all these lesions are often complicated by secondary bacterial infections (*E. coli*).

In layers it may cause reduced laying or mild conjunctivitis or may be subclinical.

- Clinical signs:
  1. Nasal discharge, sneezing, breath sounds, conjunctivitis
  2. Lot depression, quietness and clustering around heat sources
  3. Infraorbital sinusitis, exudation, open-beak breathing (Figs. 6.16 and 6.17)
  4. Occasionally, nervous symptomatology
  5. In severe cases, prostration and death
  6. Secondary bacterial infections
  7. In layers, laying hens, laying drop
- Treatment
  1. Administration of antibiotics to treat complications secondary to infection: chloramphenicol, neomycin, enrofloxacin, erythromycin or furazolidone.
  2. Keep the park environment clean, ventilated and free of ammonia.
- Prophylaxis
  1. Sensitive to most common disinfectants.
  2. Strict hygiene measures are essential to minimise the risk of contamination.
  3. Cleaning of materials, floors, indoor parks, hatcheries, etc.
  4. Vaccination with live vaccines in broilers and primary vaccination with live vaccines in layers and revaccination of layers with inactivated vaccines.

Fig. 6.16 Infraorbital sinusitis and exudation in a 13-week-old guinea fowl

Fig. 6.17 Infraorbital sinusitis and loss of eye in a guinea fowl

Infectious Bronchitis

Infectious bronchitis is a viral disease (RNA virus of the genus *Coronavirus*) affecting birds of all ages. This disease affects mainly chickens, although viruses of the same family have been isolated in turkeys, pheasants, etc.

For guinea fowl, the incidence of this disease has not been described in depth, especially in terms of problems related to egg-laying performance.

The disease is distributed worldwide. Infectious bronchitis virus attacks the respiratory tract and the urogenital tract (Cavanagh 2005; Capua and Terregino 2009; Abro et al. 2012; Brugères and Bouzouaia 2021; Jacob and Pescatore 2022).

The economic impact of infectious bronchitis is mainly due to:

ü Decreased weight gain in fattening poultry
ü Egg production losses in layers and breeders
ü Occasional mortality in broilers, layers and breeders due to kidney damage

Fig. 6.18 Egg in a feather

- Clinical signs
    1. Coughing, sneezing, nasal and conjunctival discharge, dyspnoea
    2. Sometimes conjunctivitis and inflammation of the head
    3. Depression, ruffled feathers, decreased feed consumption
    4. In layers, decrease or drop in laying
    5. Infertility in breeders
    6. Eggs in farfara (thin, fragile shells) and deformed (Fig. 6.18)
    7. Death in cases of renal involvement and secondary bacterial complications
    8. High morbidity (can reach 100%)
    9. Low mortality (can be up to 25%)
    10. Differential diagnosis with Newcastle disease virus and laryngotracheitis virus, bacterial diseases such as infectious coryza and mycoplasmosis as well as the presence of high levels of ammonia in poultry houses
- Prophylaxis
    1. Implementation of biosecurity measures on the farm
    2. Vaccination with live vaccines in fattening poultry (first days of life)
    3. Vaccination of layers and breeders with live vaccines (first days of life) as first vaccination and then vaccination with inactivated vaccine (16–20 weeks of age)

Newcastle Disease

It is caused by avian Paramyxovirus 1, a simple RNA virus, globally distributed, causing major economic losses.

It is a notifiable disease in the EU.

It is a highly contagious disease and can be found in all types of birds, including guinea fowl, in which its importance as a carrier of the virus and the importance of vaccination against Newcastle disease to suppress the carrier state in this species has been demonstrated (Capua and Terregino 2009; Miller and Koch 2013; Zvakare et al. 2018; Suarez et al. 2020).

Five forms of the disease have been described:

1. Velogenic viscerotropic—acute and affects all ages, lesions in digestive tract
2. Velogenic neurotropic—affects all ages, respiratory and nervous signs
3. Mesogenic—the least pathogenic, mortality only in young birds
4. Hitchner form—caused by lentogenic vaccinal strains, mild or absent respiratory symptoms
5. Enteric form—asymptomatic, caused by lentogenic strains that do not cause pathology or clinical signs

- Clinical signs
  Velogenic strains:

  1. Facial oedema, conjunctivitis, dyspnoea, bloody diarrhoea, paralysis, death (young animals)
  2. Sudden death
  3. High mortality (90–100%)
  4. In recovered birds, residual nervous signs

  Mesogenic strains:

  1. Respiratory signs, followed by nervous signs
  2. In laying hens, severe drop in laying and lower egg quality
  3. Mortality up to 75%

  Lentogenic strains:

  1. Mild respiratory signs
  2. Falling of the clutch
  3. Full recovery in weeks

- Prophylaxis

Since there is no treatment to prevent it, the following is recommended:

Sanitary management measures

v In the event of the declaration of an outbreak:

1. Isolate birds from the holding.
2. Limit the movement of personnel, vehicles and animals belonging to the farm.
3. Slaughter of all birds on the holding.
4. Destruction of carcasses, eggs and waste.
5. Cleaning and disinfection of facilities.

Routine management measures

1. Establish an adequate plan for dis-insectisation, rat extermination and dis-infection.
2. Do not mix batches of different ages.
3. Respect the health gap time before introducing a new batch into the rearing facilities (21 days minimum).
4. Cleaning and disinfection of farm equipment.
5. Control of visitors and vehicles.
6. Quarantine new birds arriving on the farm.
7. Avoid contact with wild animals.

Medical prophylaxis

Vaccination according to the regulations of the relevant countries.

Marek's Disease

A process caused by a highly contagious DNA-type virus of the *Herpesviridae* family, whose infections with virulent strains can cause major problems on farms. It is transmitted horizontally; there is no vertical transmission (Coles 2009).

- Clinical signs
  Classical or nervous form

  1. Paralysis of wings, legs, torticollis, respiratory problems, etc., depending on the group of peripheral nerves affected.
  2. Thickening of peripheral nerves
  3. Tumours in feather follicles
  4. Lymphomas
  5. Depigmentation of the iris (grey eye) and irregular pupil
  6. Mortality approximately 15%

  Acute visceral form

  1. Lymphomas in the viscera
  2. Depression, anorexia and diarrhoea
  3. Mortality of up to 70%

  Acute cytolytic form

  1. Highly virulent strains
  2. Lymphoid organ atrophy
  3. Very high mortality

Transient form of paralysis

1. Wing, leg and neck paralysis, lasting 24–48 hours
2. Zero mortality

- Prophylaxis

Vaccination in the hatchery on the first day of life or in ovo.

Gumboro Disease

Disease caused by an RNA virus of the *Birnaviridae* family, with two highly resistant serotypes; its most characteristic clinical sign is the enlargement of the Pouch of Fabricius (Coles 2009).

- Clinical signs:
  Acute infection

1. Develops rapidly
2. Mortality sometimes high (10–20% mortality is common)
3. High morbidity
4. Haemorrhagic lesions in thigh and pectoral muscles
5. Enlargement of the bursa of Fabricius

Subclinical infection

1. Occurs in birds with sufficient maternal immunity
2. Atrophy of the bursa
3. Immunosuppression
4. Secondary infections
5. Low production yields
6. Vaccine failures

- Prophylaxis

1. Vaccination: attenuated vaccines (primary vaccination) or inactivated vaccines (booster).
2. Separate cleaning and use of equipment for clothing, footwear, etc. per shed.
3. Installation of footbaths.
4. Eliminate beds.
5. Disinfection of parks.
6. Respect sanitary vacuum periods between batches (minimum 14 days).

Avian Influenza

Infectious process caused by an RNA virus of the *Orthomyxoviridae* family (highly pathogenic strains); it is a notifiable disease affecting most bird species and all ages; waterfowl act as reservoirs (Mo et al. 2021; Lean et al. 2022; Tweneboah et al. 2024; BehrouziNasab et al. 2025).

- Clinical signs:
  1. Respiratory symptoms: cough, sinusitis, sneezing, secretions, facial oedema
  2. Diarrhoea and nervous signs
  3. Generalised haemorrhages
  4. In layers, reduced laying
- Prophylaxis:
  1. Implement biosecurity measures on the farm.
  2. Quarantine of new birds.
  3. If an outbreak is declared:
     - Slaughter of all birds.
     - Isolation.
     - Destruction of corpses and materials.
     - Cleaning and disinfection.
     - Sanitary vacuum.
     - Establish emergency vaccination.
     - Specific measures laid down in current health legislation.

Chronic Respiratory Disease (CRD)

The causative agent is *Mycoplasma gallisepticum*, with strains of varying virulence (Jacob and Pescatore 2022; Porter Jr and Morishita 2022b).

- Clinical signs:
  1. Coughing, sneezing, eye and nasal discharges, liquid at first and then purulent, due to secondary bacterial infections.
  2. Sinusitis, aerosaculitis.
  3. Conjunctivitis (Fig. 6.19).
  4. In "pinheads", it can cause purulent inflammation of the infraorbital sinuses, giving the head a swollen appearance; sometimes, it causes permanent blindness due to involvement of the eyeball.
  5. In breeding birds, it causes a drop in laying hens and a decrease in semen production in guinea fowl.
  6. In the case of the "keets" in rearing, casualties are caused by starvation.

Fig. 6.19 Conjunctivitis in a guinea fowl cock

- Treatment
  1. Recommended treatment is with antibiotics in drinking water: tetracyclines, macrolides, thiamulin, fluoroquinolones.
- Prophylaxis
  1. Vaccination.
  2. Cleaning of parks.
  3. Improve ventilation conditions.
  4. Improve nutrition.
  5. Apply vitamin complex in drinking water.
  6. Improvement of management conditions.

Infectious Coryza

Infectious process caused by *Haemophilus paragallinarum* (Shivaprasad et al. 2022).

- Clinical signs:
  1. Conjunctivitis, sinusitis, inflammation of infraorbital sinuses (swollen head)
  2. Sneezing, head shaking
  3. Nasal and ocular discharges, liquid at first and then mucous

4. Secondary bacterial infections
5. Drop in production, in poultry, in rearing

- Treatment:
  Antibiotics in drinking water or feed: sulphonamides, quinolones and tetracyclines.
- Prophylaxis:
  1. Vaccination: bacterins
  2. Respecting the health gap periods between batches

Pathologies Arising from Handling

Pododermatitis

Infection of the limbs (usually caused by *Escherichia coli* and *Staphylococcus aureus*) with inflammation, pressure ulcers and abscesses. Occasionally, limb deformities and even loss of fingers or limb may occur.

This multifactorial disease usually starts with damage to the skin, and if the infection progresses, it can reach the muscles and even the tendons and bones.

In guinea fowl, the disease mainly affects breeding birds kept in battery cages; in particular, it is more prevalent in hens than in cockerels.

- Triggering causes of pododermatitis or "bumblefoot":
  1. Lack of exercise and sedentary lifestyles
  2. Injuries to extremities
  3. Overweight
  4. Vitamin A deficiency
  5. Lack of hygiene in the cages
- Clinical signs:
  1. Redness and swelling of the skin on the soles of the extremities
  2. The appearance of wounds and pressure ulcers (Figs. 6.20 and 6.21)
  3. Appearance of abscesses on the soles and toes of the extremities
  4. Lameness
  5. Deformities and even mutilations in the later stages of the disease
- Prophylaxis:
  The treatment of the disease in advanced stages is surgical, so in our case, it would be more effective and cost-efficient to take a series of measures to prevent the onset of this process:
  1. Revision of the extremities of the breeders.
  2. Avoid excessive weight in breeders; practice regular weight control.
  3. Provide a balanced diet.

Fig. 6.20 Plantar inflammation in the limb of a guinea fowl (claw)

Fig. 6.21 Inflammation with ulcer on the toe of a guinea fowl (claw)

4. Cleaning of the cages and the breeding room.
5. Cutting of excessively long nails.
6. Cleaning and disinfection of wounds on the soles of the extremities.

Excess Weight in Breeding Hens

Excessive feeding of breeders in battery settings causes problems of excessive fattening of the birds (Fig. 6.22), which results in a drop in laying in hens and low semen production in cockerels; in addition, this excessive weight increases the risk of heart attack deaths in birds housed in batteries (Fig. 6.23).

To avoid these situations, a strict feeding plan is necessary from the first weeks of life for guinea fowl intended for breeding, restricting the feed given to avoid excessive weight at the start of laying and semen production in the case of guinea fowl cocks.

Feather Pecking or "Cannibalism"

A process characterised by the increasing tendency of birds to peck at each other.

In the initial stages, birds appear plucked on the back and tail (Fig. 6.24); as this behaviour spreads through the flock, lesions intensify and skin wounds appear;

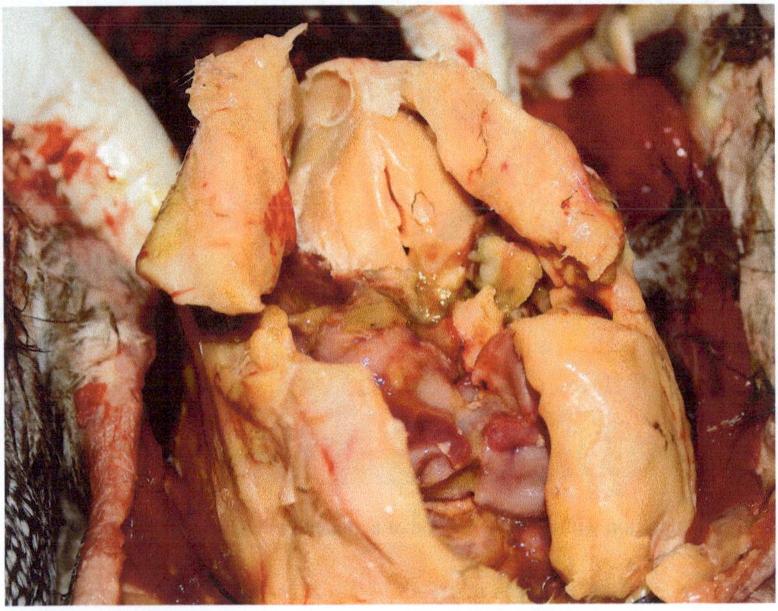

Fig. 6.22 Abdominal fat deposits in a breeding guinea fowl

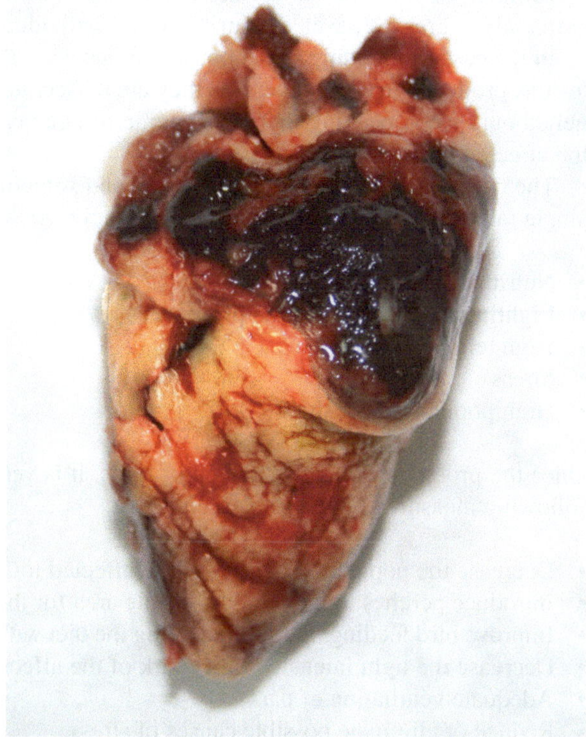

Fig. 6.23 Infarcted and greasy heart in breeding guinea fowl

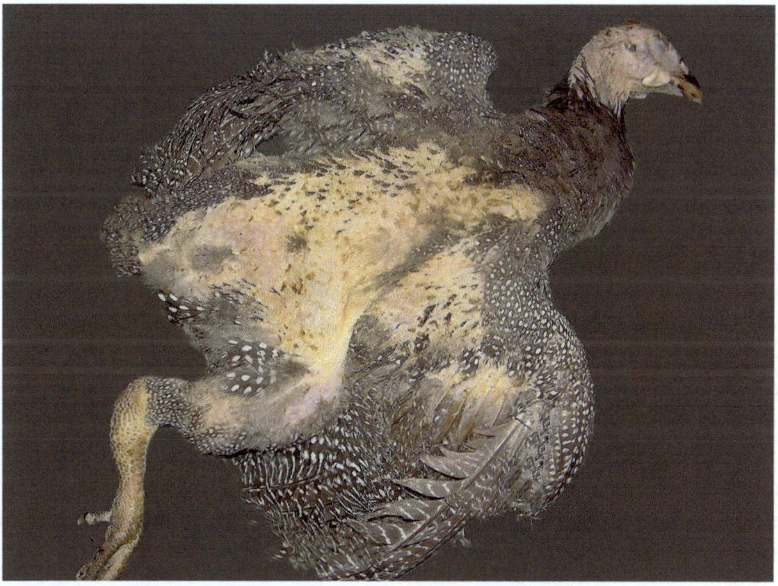

Fig. 6.24 Carcass of guinea fowl with initial signs of feather picking

bleeding from the wounds further stimulates “pecking”, until the birds become extremely aggressive, even devouring whole individuals in extreme cases.

In the case of breeding birds, housed in batteries, this process is detected by the lesions produced in some layers, as they are eviscerated through the cloaca by their neighbours when pecking them at the time of egg laying, resulting in the death of the affected bird.

The causes of this behaviour are varied and sometimes cannot be attributed to a single factor. The following events are conducive to the occurrence of “pecking”:

- Nutritional deficiencies
- Lighting too intense and too prolonged
- High temperatures
- Stress
- High population densities

Once the problem appears in a flock of birds, it is very difficult to control, but the following measures can be tried to control it:

- Decrease the population density of the affected lot.
- Introduce perches to increase the usable area for the birds.
- Improve bird feeding by supplementing the diet with a vitamin-mineral complex.
- Decrease the light intensity in the park of the affected lot.
- Adequate ventilation of parks.
- Reduce or eliminate possible causes of stress.

Dislocated Tendon

This is a chondrodystrophic condition (perosis) which is manifested by thickening of the tibio-metatarsal joint and can lead to dislocation of the gastrocnemius tendon, causing abnormal rotation of the joint. This leads to lameness, and the birds end up moving on the tarsi, resulting in inflammation (Fig. 6.25) and infection of the joint. Death is caused by starvation due to inability to reach feeders and drinkers or by injuries caused by other birds in the flock.

Some authors attribute a hereditary cause to this process, while others claim that this type of chondrodystrophy is triggered by dietary deficiencies of nicotinic acid, biotin, pyridoxine, folic acid and choline.

As a preventive measure, early strengthening of the joint bones can be attempted by giving the birds mineral salts, vitamin D3 and mineral complexes.

If we take into account the hereditary component of the process, in order to avoid an increase of cases in future flocks, it is recommended to carry out a severe “triage” of the flocks that will be future breeders, in order to avoid possible transmission to offspring.

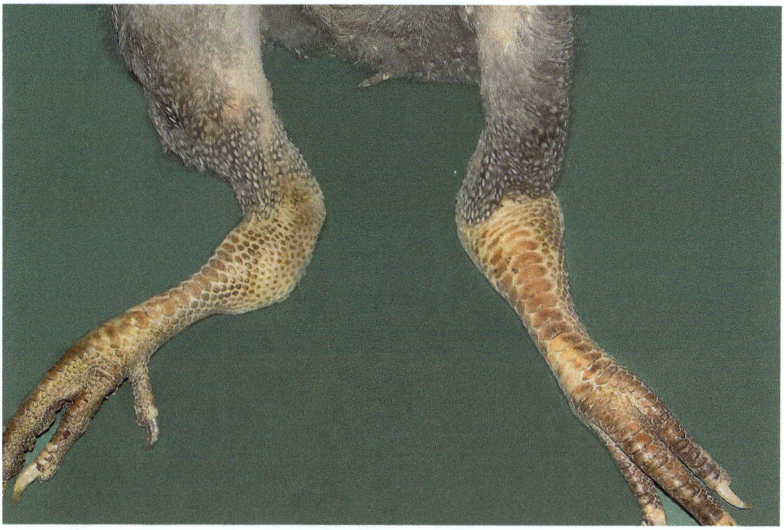

Fig. 6.25 Perosis (valgus) and joint inflammation in guinea fowl

Pathologies of Parasitic Aetiology

Coccidiosis

A parasitic disease (Fig. 6.26), whose aetiological agent is a microscopic protozoan known generically as coccidia, which includes several species of the genus *Eimeria*.

The disease affects both poultry and some domestic mammal species. In the case of poultry, it can cause multiple issues, from a decrease in the conversion rate in mild cases to high mortality rates on the farm, depending on the species of parasite and host affected. In all cases, they cause economic losses to farms; therefore, their prevention and control are very important points in any farm health programme.

One of the characteristics of coccidia is their high host specificity, which means that *Eimeria* species that parasitise some animals do not affect others, although the fact that they do not suffer from the disease does not mean that they cannot act as carriers.

The two species that most affect guinea fowl are *E. numidae* (Pellerdy, 1962) (Fig. 6.27) and *E. grenieri* (Yvoré and Aycardi, 1967) (Ortúzar-Ferreira et al. 2024) (Fig. 6.28).

Guinea fowl are relatively resistant to coccidiosis, so high mortalities can only be observed when other diseases are present. In any case, developmental delays, watery faeces, weakening and increased conversion rates occur.

Several products are used for the treatment of coccidiosis; for each of them, the correct dosage, treatment times and withdrawal periods must be applied. Some of these drugs are sulfaquinoxaline, amprolium and toltrazuril.

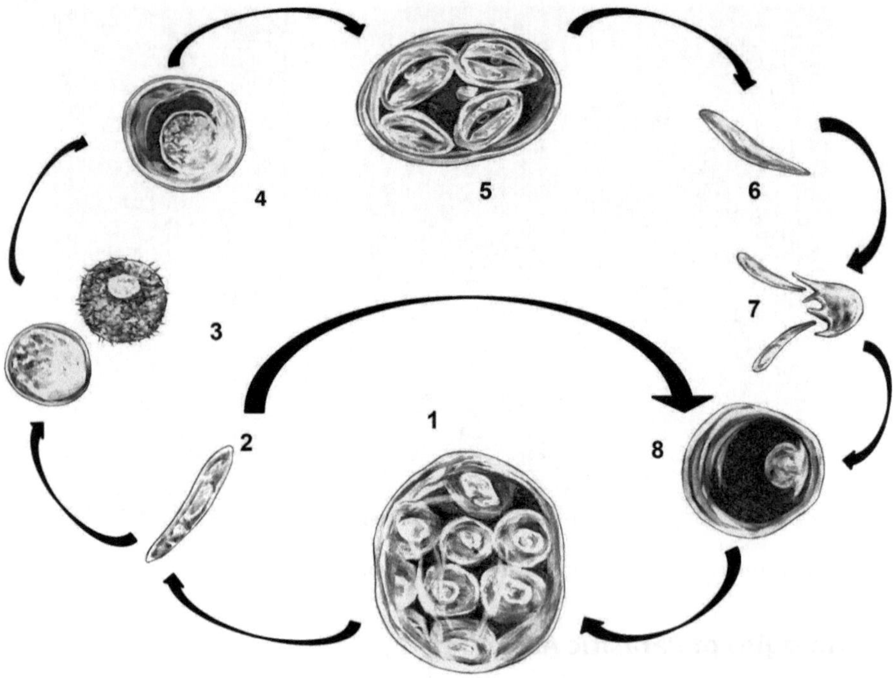

Fig. 6.26 *Eimeria* cycle. *1* Schizont with merozoites inside; *2* free merozoite; *8* formation of new trophozoite (1–2-8 ASEXUAL PHASE). *3* Formation of microgametes and macrogametes, *4* zygote, *5* oocyst, *6* sporozoite, *7* formation and maturation of the trophozoite (schizont) and release of merozoites, *8* formation of new trophozoite (1–2–3-4-5-6-7-8 SEXUAL STAGE). (Original drawing by Diego Del Moral Martínez)

Authorised preventives for Guinea fowl include amprolium, amprolium-ethopabate and methylchloropindol. Halofuginone is toxic to Guinea fowl.

Prophylactic management measures include good maintenance of litter in breeding and rearing pens, as well as avoiding any additional stress to the birds (handling, feeding or pathological) that could lead to the onset of the disease.

Trichomoniasis

This is caused by a flagellated protozoan parasite, *Trichomonas gallinae* (Rivolta, 1878) (Fig. 6.29), which inhabits the digestive tract of poultry, causing watery and foul-smelling diarrhoea, ruffled feathers, drooping wings, drowsiness and thinning, leading to irregularities in breeding flocks and reduced laying in breeders.

Lesions at caecal level are characteristic for autopsy diagnosis of the bird; when examining samples of caecal contents under the microscope, we can detect the presence of *Trichomonas* spp. by their rapid, rotating movement along the field of view of the microscope.

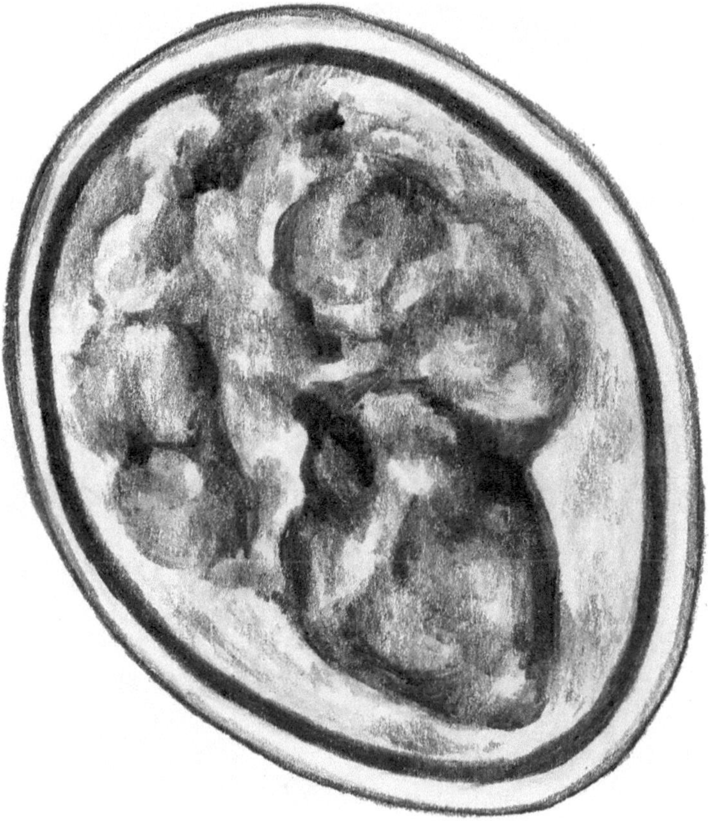

Fig. 6.27 Oocyst of *E. numidae*. (Original drawing by Diego Del Moral Martínez)

As prophylactic measures, proper litter maintenance in the parks, cleaning and disinfection of the parks, avoiding mixing of flocks of different ages and mixing of species are recommended.

Dimetridazole is the approved product for guinea fowl for trichomoniasis prophylaxis; the withdrawal period is 6 days, and treatment should be discontinued during laying.

Helminthiasis

The most frequent helminthiasis in guinea fowl is capillariasis, the two most frequent species being *Capillaria obsignata* (Madsen, 1945) (Figs. 6.30 and 6.31) in intensively reared guinea fowl and *Capillaria contorta* (Creplin, 1839) in label-produced guinea fowl (Porter Jr and Morishita 2022a).

Guinea fowl are particularly sensitive to this parasite, which causes diarrhoea, weakness, drowsiness and, in acute cases, death.

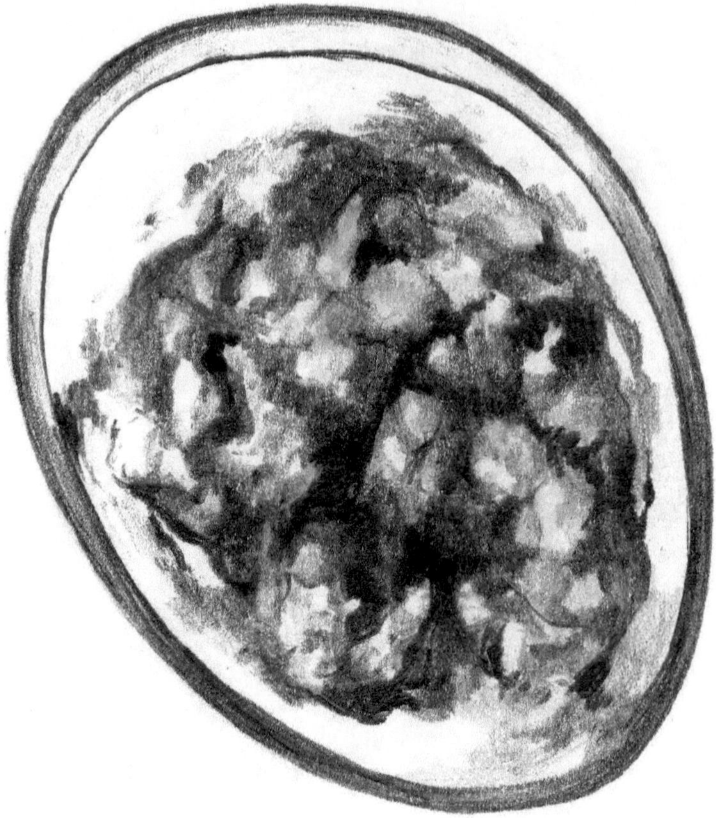

Fig. 6.28 Oocyst of *E. grenieri.* (Original drawing by Diego Del Moral Martínez)

To prevent this parasitosis, it is necessary to maintain good hygiene and management conditions on the farm; it is advisable to keep the farms well ventilated to avoid excess humidity in the litter; keeping the litter dry, frequently renewing the litter and concreting the floor will help prevent the disease. Fenbendazole can be used as a preventive and treatment.

Bird Lice

Lice affecting birds belong to the order Mallophaga chewing lice or bird lice (Fig. 6.32).

Approximately there are more than 40 species of lice in domestic birds. They are small, dorso-ventrally flat insects, rarely longer than 5 millimetres, short antennae with three to five segments, wingless and yellow or grey in colour. They are not usually haematophagous parasites but have chewing mandibles located ventrally on the

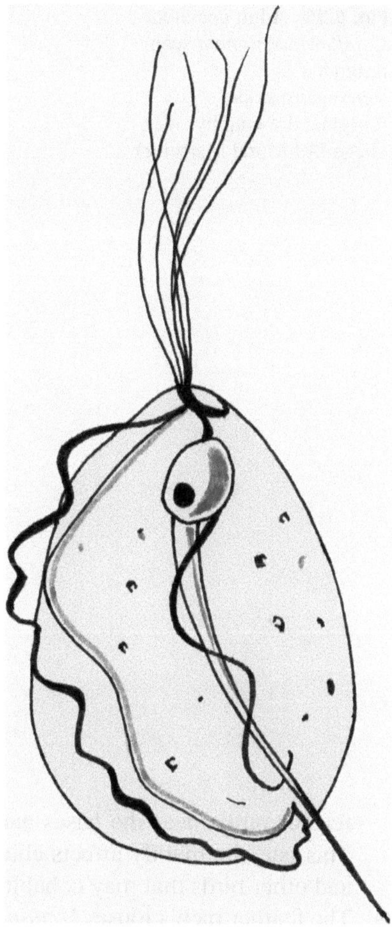

Fig. 6.29 *Trichomonas gallinarum*, flagellated protozoan. (Original drawing by Diego Del Moral Martínez)

head. They feed on pieces of feather tissue or skin, causing irritation and nervousness in birds.

They undergo incomplete metamorphosis, and their entire life cycle, including the egg stage, takes place on the body of the host.

Lice eggs are laid in clusters on the base of the feathers; in heavily infested birds, large egg masses can be found on the feathers below the cloaca.

The life cycle lasts about 3 weeks.

Some of the species described in farmland birds are as follows (Porter Jr and Morishita 2022b):

- Hen body louse, *Menacanthus stramineus* (Nitzsch, 1818) (Fig. 6.33), is one of the most common external parasites in farm poultry. It is generally located in the region below the cloaca, although in heavily infested birds, it can be found on the chest and back and under the wings. It can consume blood by puncturing soft

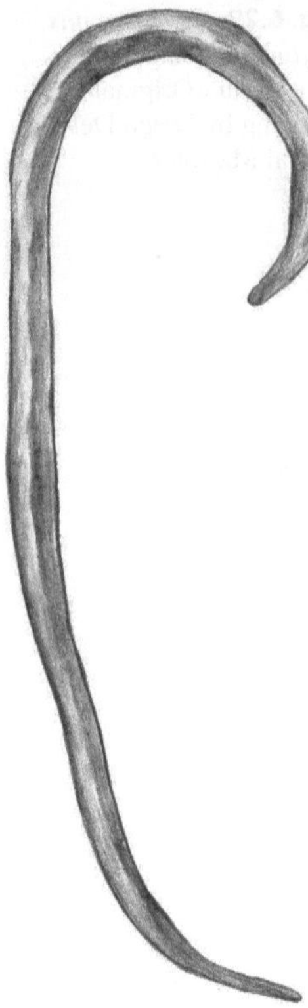

Fig. 6.30 Adult nematode *Capillaria obsignata* seen through a stereomicroscope. (Original drawing by Diego Del Moral Martínez)

feather quills near the bases and nibbling through the covering layers of skin. This parasite mainly affects chickens but may be found in turkeys, guinea fowl and other birds that may cohabit with infested hens.

- The feather rachis louse, *Menopon gallinae* (Linnaeus, 1758), is normally found along the feather rachis and does not remain on the skin of the host at any time. It is also very common on hens and guinea fowl but is of much less importance than the body louse because it spends most of its life on the feathers and not on the skin and seems to feed on the barbules and wattles of the feathers. It has not been found on young birds and may live for some time on feathers detached from the bird's body.
- The head louse, *Cuclotogaster heterographus* (Nitzsch, 1866), is found mainly on the head and neck of chickens, turkeys and guinea fowl. It is especially detrimental to young birds and is usually the only species that can cause significant damage to very young chickens, turkeys and guinea fowl.

The wing louse, *Lipeurus caponis* (Linnaeus, 1758) (Fig. 6.34), is closely related to the head louse and is the only species commonly found on the large wing feathers of gallinaceous birds. The slender turkey louse, *Oxylipeurus polytrapezius* (Burmeister, 1838), is the wing louse of turkeys and guinea fowl.

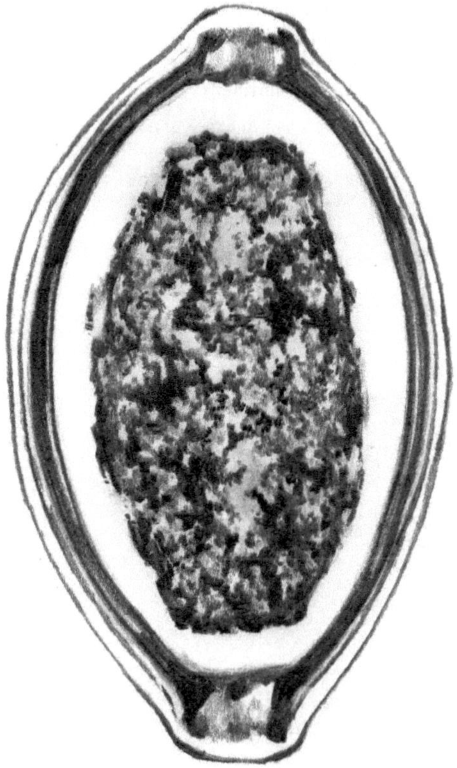

Fig. 6.31 Egg of the nematode *Capillaria obsignata* seen through a microscope. (Original drawing by Diego Del Moral Martínez)

- Other typical poultry lice belong to the genus *Goniodes* or the genus *Chelopistes* (Fig. 6.35).
- Control

  1. Pediculosis in poultry is diagnosed by finding lice on the skin or feathers of the birds. Surveillance is carried out by random and regular examination of birds in all breeding flocks and breeding rooms for the presence of lice (at least twice a month) and treatment if necessary.
  2. At least 20 to 50 birds should be examined at a time, and this should be done at random and should be chosen from all parts of the poultry house.
  3. Wild or domestic birds from outside the farm must never be allowed to come into contact with the guinea fowl. Egg-laden feathers will continue to be a source of re-infestation, and when the park is emptied, a thorough cleaning should be completed.

- Treatment

  1. Malathion as a 5% powder or 1% solution, applied to poultry, housing or flooring.
  2. Carbaryl powder 5%.

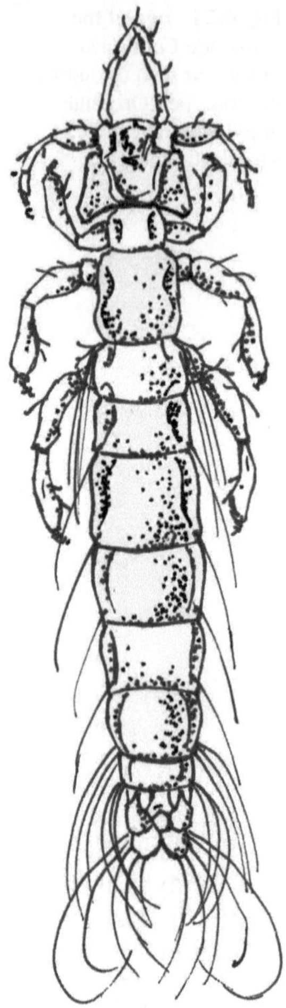

Fig. 6.32 Chewing louse, a parasite of guinea fowl. (Original drawing by Diego Del Moral Martínez)

3. 2% Imidan and Carbofenotian powder (more effective and longer lasting than malathion).
4. 4.4% Zytron granules.
5. Cage-reared hen lice can be controlled for at least 28 days by tying a resin strip containing 3.5–10% dichlorvos around one leg of each individual, at least half of the individuals, or by attaching the strips to the cage.
6. Pyrethroids (permethrin) and carbamates and their combinations are also effective, with activity lasting up to one and a half months.

Red Mite

Dermanyssus gallinae (De Geer, 1778) is an ectoparasite of poultry and other birds. It is ochre coloured and approximately 0.7 to 1 mm in size (Fig. 6.36). It feeds on

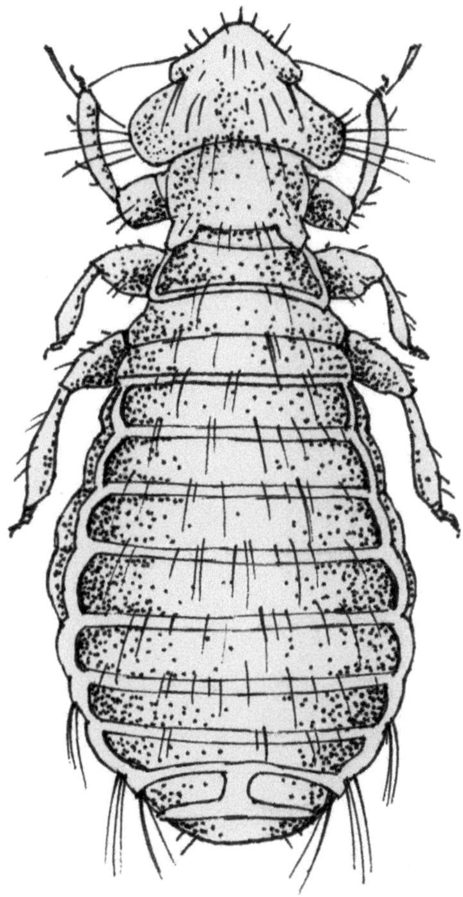

Fig. 6.33 *Menacanthus stramineus*, a common louse found on farm birds. (Original drawing by Diego Del Moral Martínez)

the blood (haematophagous) of parasitised birds. It has a life cycle of about 7 days. Eggs are laid in the shelters used for hiding.

One characteristic of this parasite is that it feeds during the night, hiding in the cracks or crevices of the breeding or rearing park during the day. They are highly resistant and can survive up to 10 months without food.

In heavily infested birds, anaemia and severe skin problems can occur (Porter Jr and Morishita 2022b).

- Treatment

  Various organophosphates, carbamates, pyrethroids and pyrethrins are used to control red mite, usually by high-pressure spraying.
- Prophylaxis

  Avoid contact with wild birds and disinfect farm equipment.

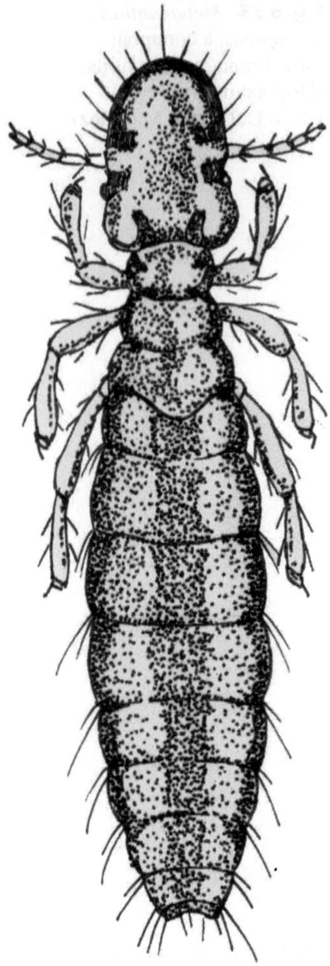

Fig. 6.34 Wing louse, *Lipeurus caponis.* (Original drawing by Diego Del Moral Martínez)

General Measures Recommended for the Health of the Farm

In order to achieve the best performance in our farm, it is necessary to establish a series of preventive measures to avoid certain infectious processes becoming established in the farm, thus causing various problems such as:

- Disease and death of breeding or fattening poultry, resulting in economic losses
- Zoonoses
- Poor quality of productions
- Excessive use of medicines, with the consequent risk of drug residues appearing on the carcasses
- Bird discomfort

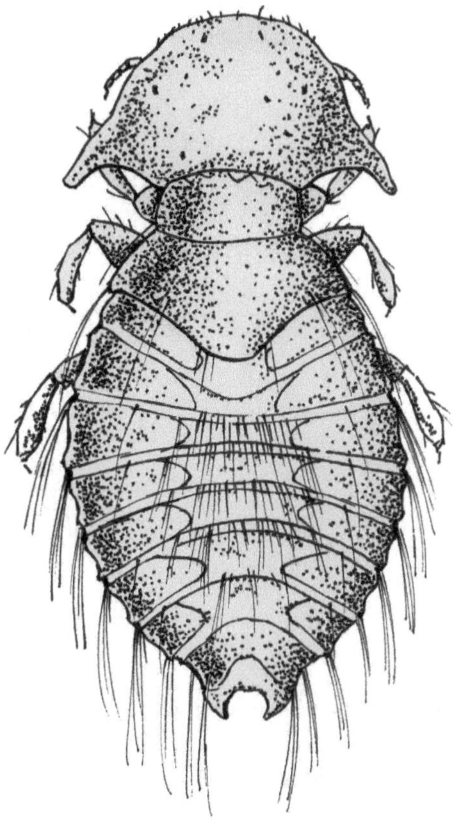

Fig. 6.35 *Goniodes*, a genus of lice commonly found on poultry. (Original drawing by Diego Del Moral Martínez)

This type of measure is used to prevent the entry of pathogens into the farm, the most important of which include the following:

- Establish a vaccination plan adapted to our farm.
- Isolate the farm from the outside (establish perimeter, separate yards, establish footbaths at the entrance of the farm or disinfect vehicle wheels with backpack and appropriate disinfectant, install bird netting on windows, install footbaths for shoes at the entrance of the shed).
- Set up the holding at the legally required distance from other poultry holdings.
- Establish an "all in, all out" management system.
- Control of unauthorised personnel.
- Use of protective clothing and footwear.
- Control of rodents, wild birds and domestic animals.
- Establish a disinfection, dis-insectisation and rat extermination programme.
- Perform sanitary vacuums and park rotations.
- Microbiological and chemical control of drinking water.
- Establish disposal system for carcasses, eggs, waste and faecal water.

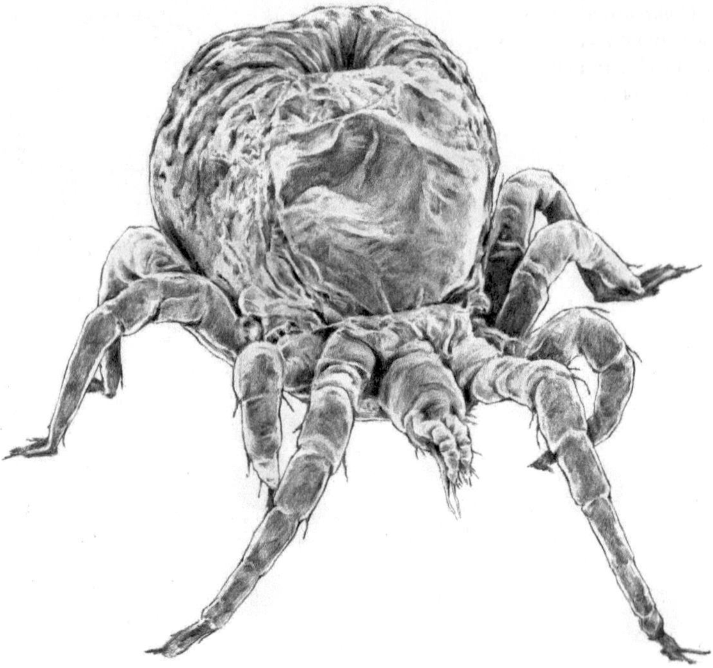

Fig. 6.36 *Dermanyssus gallinae*, red bird mite. (Original drawing by Diego Del Moral Martínez)

Vaccinations

Vaccination prophylaxis in poultry farms is essential for good results. Vaccination schedules vary significantly, as well as the type of vaccines used, because in order to establish a vaccination programme, we must take into account factors related to the geographical situation of the poultry farm, prevalence of diseases in the area, administration regulations, type of field strains present, etc.; therefore, there is no single vaccination programme.

Moreover, the more effective the on-farm sanitary measures are, the simpler the vaccination plans should be (Wambura and Kataga 2011; Jacob and Pescatore 2022; Porter Jr and Morishita 2022a, b; Abdelaziz et al. 2024).

Vaccination methods are diverse; some are described here, distinguishing between mass and individual vaccination methods.

Mass Vaccination Methods

- Vaccination in drinking water

To perform this technique, the drinkers of the birds to be vaccinated must be cleaned with clean water that does not contain products or additives that can diminish or even nullify the effectiveness of the vaccine. Residues of disinfectants must be removed, as they can damage the weakened vaccine virus.

The amount of water to be consumed by the birds to be vaccinated during the time the vaccine is to be administered must be determined. It is recommended that the water used to transport the vaccine be consumed within two hours, so that the vaccine is dissolved in this amount of water according to the number of birds to be vaccinated.

The use of stabilisers for the vaccine is also advised; one of them can be lean milk powder (25 g per 10 litres of water). The best time to perform this vaccination technique is early in the morning.

- Sprinkling

This technique allows a large number of birds to be vaccinated and reduces the stress caused to them. It consists of using a fixed or knapsack sprinkler device that allows the flow of droplets towards the birds, the size of the droplet and the distance at which the birds must approach in order to be effectively vaccinated to be regulated. To control the effective vaccination of the birds, we can observe the movement of the birds' heads when the liquid droplets that carry the vaccine reach them.

- Sprinkling inside the incubator

The chicks are placed in a spray chamber; a spray cone on the roof of the chamber discharges the vaccine into the boxes containing the birds. The distribution of the vaccine on the birds can be checked by means of a vegetable dye. It makes it possible to vaccinate a large number of birds in a short time.

- Individual vaccination methods

  1. Nasal or ocular instillation
     This method is mainly used for the start of vaccination programmes; it consists of immobilising individual birds and instilling a drop of the vaccine solution into the conjunctivae of the eyes or nasal mucosa to be absorbed through them. It generates great stress to the birds, so it is recommended for flocks of a limited number of birds.
  2. Intra-wing
     This is the method of choice for vaccination against fowl pox or smallpox encephalomyelitis. It is carried out by lancing the bird's wing membrane.
  3. Intramuscular or subcutaneous injection:
     In the case of intramuscular vaccination, this is carried out in the breast or thigh area and subcutaneous vaccination in the area of the middle third of the neck, in the area of the pygostyle. It can be carried out in the hatchery or on breeders and layers.
  4. Percutaneous or transcutaneous
     This technique does not use a needle; it deposits the vaccine dose between the subcutaneous and muscular tissues. It avoids horizontal transmission of infectious diseases. It uses a high pressure gun, applied to the breast or thigh area.

Table 6.3 Vaccination programme for broilers for rearing

Age in days	Vaccine
1	Infectious bronchitis (H120) Marek
17	Gumboro Infectious bronchitis

Table 6.4 Vaccination programme for selection and multiplication farms

Age	Vaccine
1 day	Marek
3 weeks	Newcastle (B1) Infectious bronchitis (H120) Marek
8 weeks	Newcastle (La Sota) Infectious bronchitis
10 weeks	Gumboro CRD
14 weeks	Encephalomyelitis
16 weeks	Infectious bronchitis + Newcastle + Gumboro CRD

Table 6.5 Vaccination programme for laying farms

Age	Vaccine
1 day	Marek
3 weeks	Newcastle (B1) Infectious bronchitis (H120)
7 weeks	Newcastle (B1) Infectious bronchitis (H120)
10 weeks	Encephalomyelitis
14 weeks	Infectious bronchitis + Newcastle + laying drop syndrome

In the case of meat production farms (rearing of guinea fowl), the vaccination schedule shown in Table 6.3 can be adapted.

For breeding and multiplication farms, the vaccination schedule can be as shown in Table 6.4.

A vaccination programme in layer farms is shown in Table 6.5.

Biosecurity Measures

The main resource available to a livestock producer to avoid health problems on the farm is prevention; to this end, it is essential to establish a health programme and biosecurity measures that will help minimise risks, rules which are laid down in official regulations and models of "good practice" which are described in detail in the following chapter.

Competing Interests The author has no conflicts of interest to declare that are relevant to the content of this chapter.

References

Abdelaziz K, Helmy YA, Yitbarek A, Hodgins DC, Sharafeldin TA, Selim MS (2024) Advances in poultry vaccines: leveraging biotechnology for improving vaccine development, stability, and delivery. Vaccine 12(2):134

Abro SH, Renström LH, Ullman K, Isaksson M, Zohari S, Jansson DS, Belák S, Baule C (2012) Emergence of novel strains of avian infectious bronchitis virus in Sweden. Vet Microbiol 155(2–4):237–246

BehrouziNasab O, Rouygari M, Tabari T (2025) Comparative study of vaccine-induced humoral immune response against Newcastle disease virus and avian influenza virus (H9N2) in different bird species of Mashhad birds garden. Vet Med Sci 11(3):e70191

Brugères J, Bouzouaia M (2021) Les coronaviroses aviaires. Bull Acad Vet Fr 174(1):95–101

Capua I, Terregino C (2009) Clinical traits and pathology of avian influenza infections, guidelines for farm visit and differential diagnosis. In: Avian influenza and Newcastle disease: a field and laboratory manual. Springer Milan, Milano, pp 45–71

Cavanagh D (2005) Coronaviruses in poultry and other birds. Avian Pathol 34(6):439–448

Coles BH (2009) Galliformes. In: Handbook of avian medicine, p 309

Cruciani D, Crotti S, Maresca C, Pecorelli I, Verdini E, Rodolfi M, Scoccia E, Spina S, Valentini A, Agnetti F (2022) Preliminary investigation about *Aspergillus* spp. spread in Umbrian avian farms. J Fungi 8(11):1213

Hoffman G, Völker H (1969) Anatomy and physiology of domestic fowl. Ed. Acribia, Zaragoza, Spain, pp 160–164

Jacob J, Pescatore A (2022) Guinea fowl. In: Gamebird medicine and management, pp 297–318

Le Coz-Douin J (1992) L'élevage de la pintade. Editions du Point Vétérinaire, Maisons-Alfort, France

Lean FZ, Vitores AG, Reid SM, Banyard AC, Brown IH, Núñez A, Hansen RD (2022) Gross pathology of high pathogenicity avian influenza virus H5N1 2021–2022 epizootic in naturally infected birds in the United Kingdom. One Health 14:100392

MAPA (2004) Ministry of Agriculture, Fisheries and Food/Inprovo. Guide of good hygiene practices in laying poultry farms. Madrid (Spain)

MAPA (2005) Ministry of Agriculture, Fisheries and Food. Guide of good hygiene practices in poultry breeding farms for the control and prevention of zoonotic salmonellas. Madrid (Spain)

McMullin, P. F. (2020). Diseases of poultry 14: David E. Swayne, Martine Boulianne, Catherine M. Logue, Larry R. McDougald, Venugopal Nair, David L. Suarez, Sjaak de Wit, Tom Grimes, Deirdre Johnson, Michelle Kromm, Teguh Yodiantara Prajitno, Ian Rubinoff & Guillermo Zavala (Eds.), Hoboken, Wiley, 1451.

Miller PJ, Koch G (2013) Newcastle disease Diseases of poultry, vol 13, pp 89–138

Mo J, Youk S, Pantin-Jackwood MJ, Suarez DL, Lee DH, Killian ML, Bergeson NH, Spackman E (2021) The pathogenicity and transmission of live bird market H2N2 avian influenza viruses in chickens, Pekin ducks, and Guinea fowl. Vet Microbiol 260:109180

Ortúzar-Ferreira CN, Dorna-Santos L, de Oliveira MS, de Lima VM, Duszynski DW, Berto BP (2024) Coccidia of Guinea fowls: validity of recorded *Eimeria* spp. (Apicomplexa: Eimeriidae) and first molecular identification of *Eimeria grenieri* Yvoré & Aycardi, 1967. Parasitol Int 103:102937

Porter RE Jr, Morishita TY (2022a) Gamebird digestive diseases. In: Gamebird medicine and management, pp 152–195

Porter RE Jr, Morishita TY (2022b) Gamebird skin diseases and multisystemic diseases. In: Gamebird medicine and management, pp 214–244

Shivaprasad HL, Montiani-Ferreira F, Moore BA (2022) Ophthalmology of galloanserae: fowl, waterfowl, & relatives. In: Wild and exotic animal ophthalmology: volume 1: invertebrates, fishes, amphibians, reptiles, and birds. Springer International Publishing, Cham, pp 571–626

Sisson S, Grossman JD (1982) Anatomy of domestic animals, 5th edn. Ed. Salvat, Barcelona, Spain, pp 1955–1959

Sturkie PD (1968) Avian physiology. Editorial acribia, Zaragoza-Spain, pp 152–154

Suarez DL, Miller PJ, Koch G, Mundt E, Rautenschlein S (2020) Newcastle disease, other avian paramyxoviruses, and avian metapneumovirus infections. In: Diseases of poultry, pp 109–166

Tweneboah AA, Johnson SAM, Amponsah PM, Asare DA, Emikpe BO (2024) Seroprevalence of avian influenza in Guinea fowls in some districts in the upper east region of Ghana. Vet Med Sci 10(6):e70106

Ugochukwu ICI, Aneke CI, Sani NA, Omeke JN, Anyanwu MU, Odigie AE, Onoja RI, Ocheja OB, Ugochukwu MO, Luca I, Makanju OA (2022) Important mycoses of wildlife: emphasis on etiology, epidemiology, diagnosis, and pathology-a review: part 1. Animals 12(15):1874

Wambura PN, Kataga S (2011) Putative protective antibody response following oral vaccination of multi-age free ranging helmeted Guinea fowls (*Numida meleagris*) with Newcastle disease virus strain I-2 coated on oiled rice. Trop Anim Health Prod 43:99–102

Woernle H (1994) Enfermedades de las aves Editorial Acribia. Zaragoza (Spain)

Zvakare P, Mugabe PH, Mutibvu T (2018) Guinea fowl (*Numida meleagris*) production by small-holder farmers in Zimbabwe. Trop Anim Health Prod 50:373–380

Engineering: Constructive Characteristics of Installations, Machinery and Tools

7

Cesáreo Sánchez López

Abstract

Any livestock farming project must meet the necessary requirements for the type of production it intends to carry out, which will include the necessary measures to preserve the safety of the operators, guarantee animal welfare and prevent environmental pollution. These requirements, both technical and administrative, are regulated by legislation at municipal, regional, national and supranational levels. Their description, with the specific characteristics required for the construction of guinea fowl farms, is included in this chapter dedicated to the farming of guinea fowl. In order to carry out this activity, it is also necessary to draw up a technical project and obtain health authorisation from the autonomous community/region in which it is to be carried out, which requires prior authorisations such as environmental authorisation, urban planning qualification, municipal building permit, municipal activity permit and approval of the operating report and health programme. This chapter also includes the technical characteristics of the project regarding the housing and facilities for breeders, breeding and rearing and general facilities, as well as the necessary conditions to achieve animal welfare, descriptions which are illustrated with photographs, sketches and diagrams.

Keywords

Legislation · Administration · Livestock farming · Livestock project · Guinea fowl

C. S. López (✉)
CASTRUM ERAT Farm, Castuera, Extremadura, Spain
e-mail: info@castrum-erat.com

J. del Moral de la Vega, J. del Moral Martínez (eds.), *The Guinea Fowl (Numida meleagris)*, https://doi.org/10.1007/978-3-032-11654-3_7

The Project of a Livestock Farm

The different legislations that regulate production activities, including livestock farming, are becoming increasingly demanding, as production, to the extent that different technological and genetic advances, etc. are developed, must assume an increase in production risk. This production risk, which translates into pollution and/or waste production, must be minimised so that livestock farming can continue to generate wealth and produce products without being rejected socially (Bist et al. 2024).

It is in this line of work that the legislator proposes a series of rules that regulate the different productive activities with the aim of making social demands and the satisfaction of consumers' needs compatible. For this reason, the set of existing rules tends to protect fundamentally:

- Production per se.
- Individual and collective interests.
- Territory and the exploitation of natural resources.
- Competition between production systems, in an open market on a global scale.

However, these aspects, embodied in legal texts, are often far from daily practice, in the drafting of the project, in the legalisation of a livestock farm, in the processing of documentation, etc.

The prevention of environmental impacts is the best solution to be able to develop an industrial activity, such as modern livestock farming. If the different legal aspects that affect livestock farming are not taken into account, it can lead to an unreasonable situation that will cause the project to stall.

The wide legislative framework, in volume and diversity, that affects the constructive design of livestock housing (Adams et al. 2025; Nielsen et al. 2025; Zamani et al. 2025) is difficult to bring to a common and practical synthesis, either because of the large number of provisions that affect it or because of the successive amendments and/or incorporation of new legal texts to the existing body of regulations in different nations.

The engineer, the technician who develops the design of the livestock housing or facilities, will have to be informed about the current legislation at local/municipal level, at regional level, at national level and at supranational level, which may affect the future construction or development of the livestock production activity.

At local/municipal level, a poultry farm is considered, for legal purposes, as a nuisance activity (noise and odours), unhealthy (waste generation) and harmful (possible infectious diseases) and therefore requires a municipal activity licence to operate legally.

The various procedures for obtaining a municipal/premises licence for a guinea fowl farm are usually no different from any other livestock activity, and the service of a technician specialised in the processing of livestock activities must be provided, whose first task is to ensure that the location for the installation is favourable. We will look at the factors to be taken into account in order to reach this conclusion later on.

The choice of land is very important, as it will determine, firstly, the viability of being able to develop our project and, secondly, how easy it will be to expand it in the future.

Formalities for the Legal Development of the Activity

Before proceeding with the realisation of an agricultural project, in this case, a guinea fowl farm, we must carefully study the location.

Livestock farms, zoos and other activities involving the breeding, fattening and selection of animals, as indicated above, are usually considered as nuisance, unhealthy, noxious and/or dangerous activities, and when legally regulated, depending on how the activity is considered, a minimum distance is usually established from urban centres and dense population centres, and an average reference distance of 2000 m can be established.

In order to obtain the Operating Licence for a guinea fowl farm, a Technical Project of Civil Works and Installations, signed by a competent technician, must be submitted. The competent technicians for these activities are agricultural engineers, although this may vary according to the different legislative realities.

This project will be the set of documents that define, in accordance with the regulations in force, the works to be carried out and the required installations in order to, in agreement with the promoter, seek the greatest profitability for the designed operation.

It is very important that the promoter imposes certain starting conditions, which should include the volume of production to be carried out and the investment to be made. These conditions must be clear before any steps are taken.

A very common condition imposed by the developer is the location of the farm. It is quite normal that when someone decides to set up a livestock farm, it is because in most cases, they have the land. In this sense, it is necessary to be serious and rigorous, as not all land is suitable for all purposes, and the technician in charge of the work must advise the future livestock farmer in this regard.

The first thing to consider is the distance to the urban centre and then the distance to sources of infection, such as other poultry farms, slaughterhouses, carcass centres, etc., which are usually subject to health regulations.

Special attention must be paid to watercourses, as the existing hydraulic regulations must be respected, which usually limit the construction of installations at less than one distance from the maximum flood line of the watercourse (a reference average of about 5 m can be established) and at less than another distance from the basin (a reference average of about 100 m can be established). Finally, an Urban Planning Certificate of the plot will be requested from the Town Council of the location where the activity is to be carried out. In this certificate, information will be provided on the permitted uses for this land and the urban development conditions, which must be faithfully justified in the Technical Project.

With this data, we have preliminary information to determine whether in principle it is feasible to occupy this land for the installation of our guinea fowl farm.

Our final objective is to be able to breed hens on the selected plot. For this, as for any livestock farm, a Health Authorisation is usually required from the regional administration.

In order to obtain this Health Authorisation, documentation generally has to be submitted together with the application, in its simplest form usually consisting of an Operating Report, a Health Programme and the Operating Licence.

As mentioned above, the competent body for granting the Operating Licence is usually the Town Council of the area where the activity is carried out. In order to obtain this licence, depending on the specific legislative reality, it is usually necessary to obtain other prior documentation such as a Municipal Works Licence, an Environmental Authorisation and the Urban Classification for the works and the activity.

The documentation that might be needed to obtain each of these permits or reports is discussed below.

A. Environmental Authorisation.

In order to analyse the environmental repercussions that our activity may generate, it is necessary to differentiate between three concepts.

- Environmental Communication: this is the documentation by means of which the promoter of an activity or installation informs the City Council of the data and other requirements necessary to accredit compliance with the environmental conditions and technical requirements for the implementation of activities and installations with a low environmental impact.
- Unified and/or Integrated Environmental Authorisation: this is the resolution that would be granted by the competent autonomous/regional body, which would allow, for the sole purpose of protecting the environment and human health, the operation of all or part of the industrial facilities, under certain conditions aimed at ensuring that they comply with the established purpose and the corresponding legal provisions. Whether it is unified or integrated would depend on the size of the operation.

Let us analyse the possible Unified Environmental Authorisation, as it is the most suitable for an average farm.

This authorisation integrates in a single act of administrative intervention the authorisations, mandatory sectoral reports and necessary prescriptions regarding atmospheric pollution, including determinations concerning volatile organic compounds, discharges to the integral sewerage system, waste production and management, contaminated soils, noise pollution and light pollution.

Essentially, it tends to take up the figures already existing in the different environmental regulations:

1. In the field of environmental impact assessment: the environmental impact statement or the environmental impact report.
2. On air pollution:
   - The authorisation or notification of emissions of air pollutants into the atmosphere from potentially air-polluting activities
   - Notification of certain activities emitting volatile organic compounds
3. With regard to discharges into the integral sewerage system:
   - authorisation of discharges into the sewerage networks
4. In the field of waste: the authorisation for the production of hazardous waste and the authorisation for waste management.
5. On contaminated soils: the reports and prescriptions required for contaminated soils
6. In terms of noise pollution: the noise requirements
7. Regarding light pollution: the requirements for light pollution

As indicated above, unified environmental authorisation must be obtained before the other sectoral authorisations or licences that are required for each activity, including urban planning licences (both the works, building and installation licence and the licence for uses and activities).

The application shall be addressed to the competent autonomous/regional body, accompanied by the documentation suggested below:

- Unified environmental authorisation application form.
- Basic project, signed by a competent technician, which develops the information relating to the environmental aspects that are the object of the unified environmental authorisation.

These include the following:

1. Background
   1.1 Introduction
   1.2 Purpose of the project
   1.3 Owner of the installation
   1.4 Location: municipal district, industrial estate and plot; access to the facilities; geographical and UTM coordinates; distances
   1.5 Applicable legislation
2. Activity and facilities:
   2.1 Description and classification of the activity; type of activity and production capacity; description of the production cycle; compliance with technical-sanitary regulations; farming system (management plan, housing system, feeding system, drinking system, ventilation system, heating system, other elements of the operating system); timetable for implementation and commissioning

   2.2 Description of the installations, supported, where appropriate, by graphical information; main and auxiliary; environmental; sanitary; other infrastructures; supply system, efficiency measures and expected annual consumption of resources (water, energy and fuels, feed, chemicals); description of the machinery used and its characteristics
3. Waste and by-product management:
   3.1 Animal health, plant health, veterinary drugs, feed additives and other wastes
   3.2 Management of carcasses
4. Emissions to water, soil and air:
   4.1 Air emissions; diffuse livestock emissions—best available techniques to replace emission limit values; other activity-related emissions
   4.2 Emissions to water
   4.3 Noise emission data (including activities in parallel with the operation)
5. Slurry, manure and poultry manure management programme:
   5.1 Slurry and manure production
   5.2 Extraction system from the pit/pond or tank, frequency of emptying
   5.3 Destination; Manure Agricultural Application Plan; delivery plan and identification of the manager (in the event that all or part of the manure produced is delivered to a manager for treatment and/or application)
6. Budget.
7. Plans:
   7.1 Topographical location
   7.2 Georeferencing of buildings and facilities
   7.3 Plant, equipment and infrastructure

- Report from the Town Council in whose municipality the installation is located, accrediting the compatibility of the project with urban planning.
- Environmental impact study or environmental document.
- Documentation that identifies and accredits the ownership of the installation, providing the documentation that is applicable in each case: deeds of incorporation of the property, powers of attorney, deeds of ownership of the land or lease contract, individual identification documentation of the person signing the application.
- Non-technical summary of all the indications specified in the previous points, in order to facilitate their understanding for the purposes of the public information procedure.
- Identification of the data which, in the applicant's opinion, are confidential in accordance with the provisions in force, with an express indication of the legal regulations that protect such confidentiality.

B. Urban Classification

This type of activity must, of course, be carried out on rural land.

Depending on the area, there may be different regulations governing land use and conditions of use.

The urban planning indices that can most condition the operation are the minimum plot area suitable for building, the distance to boundaries and the buildability of the plot. Here we can find different values between the autonomous/regional law and the subsidiary norms of the municipality to work with. In any case, we will always be obliged to comply with the most restrictive conditions.

It is very important to check that our plot allows the uses we intend to develop on it (Dupas et al. 2024). To achieve this, the most advisable thing to do, as indicated above, is to request the Urban Planning Certificate of the plot under study before starting any phase of the project. This document will inform us of all the urban planning conditions of the area: minimum plot area required, separation from boundaries, building heights, buildability, etc. and the uses, which in our case—to be able to carry out a farm of guinea fowl—will be livestock activity.

Livestock activity is usually considered as a primary and natural use of the land when it is used extensively. In our case, as we are dealing with a livestock farm with facilities to house breeding animals, breeding sheds, etc., it is necessary to develop the urban planning qualification of the works and the activity.

In order to be able to proceed with the Urbanistic Qualification, a Technical Report must be prepared for which the following documentation is suggested:

- Justification of the suitability of the works for the environment in which they are located.
- Area occupied by the building and description of its basic characteristics, indicating:
  - Occupied floor area
  - Total built-up area
  - Number of floors
  - Construction height
  - Type of cover
  - Exterior walls
  - Water supply system
  - Sanitation system, indicating destination and treatment
- Distance from the building:
  - Boundaries
  - Urban core
  - Buildings within a 300 m radius
  - Highways and roads
  - Maximum flood line, if applicable
- Justification that there is no possibility of the formation of a population nucleus.
- Justification that, for the location of the same, there is no other suitable land with suitable urban qualification for the intended use.

- If there are other buildings on the property, please indicate for each of them:
  - Typology and use
  - Occupied area
  - Total existing built-up area
  - Number of floors
  - Authorised building permit or application for legalisation of the works
- Situation plan of the plot, at a minimum scale of 1:10,000 with respect to one of the plans in force.
- Plan of the building within the plot, indicating the distance to boundaries and the area of the rustic unit suitable for building.
- Plan, elevation and cross-section drawings of the building.

The following documentation should also be attached to this Technical Report:

- Proposal, by the developer, of the free transfer of land corresponding to the municipality as a participation in the urban development granted by the qualification or, where appropriate and when so accepted by the municipality, the amount of the substitute fee to be paid to the municipality.
- Acceptance, on the part of the City Council, of the transfer of land corresponding to the municipality as urban use granted by the qualification or, where appropriate, the amount of the canon in lieu of this transfer.
- Proof of the surface area of the property.
- Environmental impact study, which should indicate reforestation proposal and restoration plan.
- Sectoral reports required by reason of their location or activity.

C. Municipal Works Licence

When the Technical Project of our farm is presented to the Town Hall, a Municipal Works Licence is requested. This will be granted after a study of the Technical Project by the municipal technician who will check that it complies with urban and environmental regulations. For this, we would have to have authorised the Urban Classification of the works and activity and the Unified Environmental Authorisation.

At this stage, the works can begin, but not before paying the corresponding fees for a Municipal Works Licence, which is a percentage of the final budget for the material execution of the project, which is usually between 2% and 5%, and proceeding to the free transfer of land corresponding to the municipality, which must be at least 15% of the land linked to the qualification, or if applicable, and when the municipality has accepted it, the amount of the substitute fee to be paid to the municipality.

It must be clear to the developer that before starting to operate, he must first start paying taxes, which is no different from any other investment activity to be carried out on rural or urban land.

D. Municipal Operating Licence

This is usually granted together with the Municipal Works Licence, and as its name suggests, it authorises the activity to be carried out, in our case, the exploitation of a poultry farm.

For this purpose, the municipal technicians will have obtained the relevant reports from the Health Department, regarding distances from the farm to urban centres or grouped population centres, points of infection and measures to control the appearance and spread of diseases.

With this licence granted, all that remains is the Health Authorisation to be able to start work.

E. Health Authorisation

As it is a livestock activity, we cannot forget the sanitary conditions imposed by the central, regional and local administrations.

A veterinarian responsible for the farm shall draw up a health programme containing at least the following points:

- Define husbandry technique: As far as possible, husbandry technique shall be based on the principles of protected husbandry and "all in, all out", within each house. Between each flock, cleaning, disinfection and depopulation shall be carried out.
- Define the origin of consignments of animals or eggs to be hatched.
- Establish hygiene standards to be observed by farm staff and visitors.
- Describe cleaning, disinfection, dis-insectisation and rat extermination programmes, how and by whom they are carried out and also the products used for these tasks.
- Express the origin of the feed (purchased and/or self-consumption).
- Define the source of drinking water and chlorination programmes.
- Describe the health programme carried out on the farm, defining the poultry diseases for which special surveillance is to be carried out.
- In the case of hatcheries (operation based on the principle of one-way circulation of eggs, operating equipment and personnel), the microbiological quality control programme shall be described.

The health programme for the control of infectious-contagious and parasitic diseases must be carried out on all holdings.

In order to register the farm in the possible existing Farm Register and to obtain the Health Authorisation, a Technical Veterinary Report, signed by the veterinarian of the farm, must be submitted, which should at least contain the following aspects:

- Situation of the holding:
  - Municipal area
  - Sketch of the situation of the holding
  - Polygons and plots
- Type of poultry farm applying for:
  - Reproduction-selection.
  - Reproduction-multiplication
  - Breeding poultry hatchery
  - Breeding of laying hens
  - Broiler production
  - Incubator
- Species of poultry to be kept on the holding. If more than one species is kept, it must be indicated that they are well separated from each other.
- Indicate the type of rearing in the case of laying hens. If reared in cages, indicate the type of cage.
- Number of vessels making up the holding:
  - Cubic capacity of halls in m^2
  - Maximum capacity of birds at a time per house and type of rearing
- With regard to the facilities, mention should be made of the following:
  - That the facilities prevent birds from escaping
  - That the location and layout of the facilities are appropriate to the type of production undertaken
  - That the facilities ensure good hygienic conditions
  - That the facilities allow the practice of sanitary control
  - That the materials used are suitable and allow easy cleaning and disinfection
  - That the facilities have protection against birds and rodents
  - That the light and ventilation of the facilities are adequate
  - That it has running water and the source of the water
  - That it has a lazaretto on the farm
  - Waste and carcass disposal
- In hatcheries, there must be physical and functional separation between the hatchery and other rooms on the farm, and the walls and floors of the facilities must be waterproof and easy to clean and disinfect.

What has been explained so far has referred to all the documentation and the steps to follow in order to obtain the relevant authorisations for the legal development of our activity.

Next, we will develop the aspects to be taken into account when developing the project, aspects related to the facilities, where special attention will be paid to the housing for guinea fowl rearing, its orientation and ventilation and a review of the requirements in terms of animal welfare (Dobraszczyk 2024).

Accommodation and Facilities

Basic Considerations

Before going into detail on the particularities of the housing and equipment that we consider necessary for rearing guinea fowl, it is necessary to refer to the economic constraints that are generally imposed by the small volume of production achieved by the vast majority of farms.

The family character of many farms, born with the purpose of complementing the domestic economy and, also, the lack of knowledge of the needs of the birds to be farmed and of the environmental factors that can affect their zootechnical performance, also contributes to the tendency towards simple constructions, with no or low technical quality and towards the use of existing buildings without any type of remodelling or adaptation to the necessary requirements.

While it is important, and every effort should be made to economise on investments and to try to make use of existing premises, it is no less important that this should be done on the basis of a minimum knowledge of the vital needs of the bird to be reared, at every stage of its life: space, possible animal density, ventilation, heating, cooling, etc.

The basic principles applicable to the design of facilities for guinea fowl are generally common and can be extrapolated to other poultry species, such as game birds or ostriches (Fig. 7.1).

We consider the basic infrastructure of the farm where it is to be set up to be the availability of water, electricity and adequate access.

Water is a basic element that makes it possible to run the farm. Water is the drink for the poultry, but it is also necessary for cleaning the facilities and equipment.

Water has to be considered as an irreplaceable nutrient for poultry. Its quality influences the yields per bird, so it is important to monitor this aspect on a regular basis.

The required volume of water should be calculated in advance to ensure that it is available at all times. On average, a guinea fowl consumes about 0.350 l per day. The cleaning of rooms and equipment is generally calculated at 500 l of water per 100 m3 of room.

In addition to ensuring the water supply, the availability of water at any time shall be ensured, anticipating the possibility of interruption or restriction due to breakdowns or drought. For this purpose, large-capacity tanks or cisterns shall be provided to guarantee drinking water for the hens for a few days.

Electricity is necessary for lighting installations and for the operation of certain equipment. But its availability has an easier solution than that of water in those places far from the electricity grid, for which the connection can be an important part of the investment to be made. In such cases, the installation of generators of greater or lesser productive capacity can solve the problem.

In general, poultry farms do not require significant electrical power, unless there are, at the same time, the lighting of birds, incubation of a certain volume, cooling of premises, use of motors, etc.

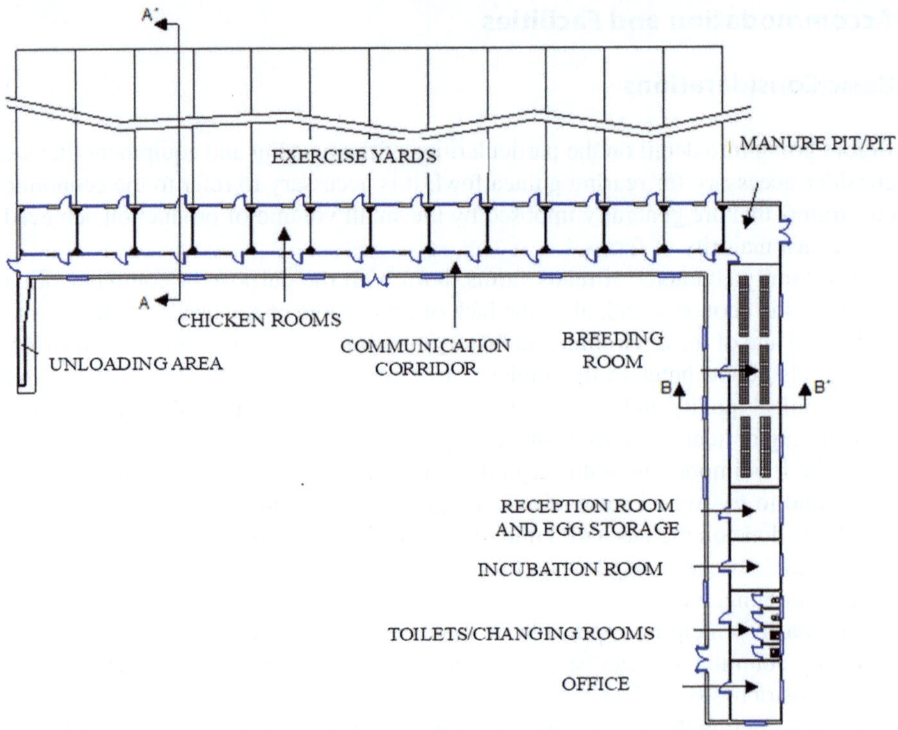

Fig. 7.1 Typical layout of an extensive laying farm

Lighting of the birds is generally limited to the first days or at most the first weeks of age of the guinea fowl, with intensities of 2 a 10 W/m^2. This lighting is also the heating applied to the young chicks.

Access to the farm should be easy and comfortable, without any accidents that could make the way to the farm treacherous. If necessary, access should be adapted to the traffic of heavy vehicles supplying feed or transporting the hens to the slaughterhouse.

The entrance to the farm must always be restricted by means of signs, fences or barriers, for simple principles of sanitary and qualitative safety. At the entrance it is advisable to build a sanitary ford to disinfect vehicles.

When it is possible to choose between several poultry farms, it is worth taking the time to select the most suitable one, depending on the poultry production you wish to undertake. Apart from the orographic and climatic considerations, it is also necessary to consider the compulsory distances to urban centres, other poultry farms, watercourses, etc. mentioned in the previous point.

It is important to consider the noise level in the area where the farm is to be located, as it can disturb the harmony and peace of our birds. Do not forget that our birds also produce noise, and we will have to evaluate what problems this may cause.

Fig. 7.2 It is highly recommended that the exercise yards have grass, reseeding the most suitable species during the period when the yards are empty

A strategic location must be sought with respect to suppliers of feed, medicines, fuel, slaughterhouses, specialised veterinary surgeons, road network, etc.

Preferably, flat, but slightly sloping land should be chosen for the location of the farm with its annexed parks, so that rainwater can run off. Care should be taken to prevent the formation of pools where birds can drink. This water will be contaminated by chicken droppings, algae and numerous microorganisms.

In general, land with a slope of no more than 5% is preferable, as it also makes it easier for farm staff to manage.

The buildings (initial rearing premises, poultry houses, feed storage, equipment, etc.) should be located in the higher areas, as they are the most airy, sunny and dry. Parks shall be built on one or both sides of it.

Parks should be planted with vegetation, preserving native vegetation or sowing some cereals or meadows (Fig. 7.2).

Fences should be constructed of strong wire mesh, supported by posts and buried underground to about 30 cm.

Accommodation

Livestock housing and facilities have become a very important (but certainly not sufficient) necessary element in modern livestock production models.

Its importance generally increases in direct proportion to the level of technical and economic intensification that characterises the production model (in this sense, the level of demand of a broiler chicken is very different from that of a Castilian hen).

Professor Buxade (Castelló 1996; Buxade 1998) establishes the following aspects when designing the farm:

- Animal base to be exploited.

Obviously, the anatomical-physiological characteristics of the animals to be kept on the farm will be decisive in establishing the requirements in terms of actual usable area per production unit, environmental conditions, etc.

- Operating system.

From the outset, it will influence the design of our installations, as the design of installations for a modern free-range hen house will be completely different from that for a battery hen house with inverted cages.

In general, accommodation should be characterised by:

- Functionality. A good design of the accommodation and facilities allows for greater rationalisation and greater ergonomics of the work to be carried out, especially those that are carried out more frequently, and which are also, logically because of their repetition, those that give rise to errors, inefficiencies and/or ineffectiveness.
- Location and orientation. Consideration should be given to prevailing winds, landscape impact, terrain conditions, etc.
- Adequate overall conception. The versatility of the facilities should be sought in their design, their capacity to adapt to changes, the possibility of extensions or modernisations.
- Correct choice of materials. All materials are important, but some of them, such as roofing, insulation or flooring, must be particularly carefully considered in poultry farming. When choosing materials (apart from the question of economy and simplicity), it is necessary to look for harmony and harmony between them, taking into account, more and more every day, the external appearance of the building, which should not be luxurious, but fit for purpose.
- Profitable mechanisation. Any livestock farm (as is the case for agriculture) can, a priori, admit an investment in mechanisation capable of ruining any livestock farmer and/or being absolutely unamortisable (Fig. 7.3).
- Waste management. It is necessary to take advantage of the topographical characteristics of the land where the farm is to be located in order to design the evacuation, transfer, handling and treatment of the excrement. Special attention should be paid to dung heaps and pits or ponds.

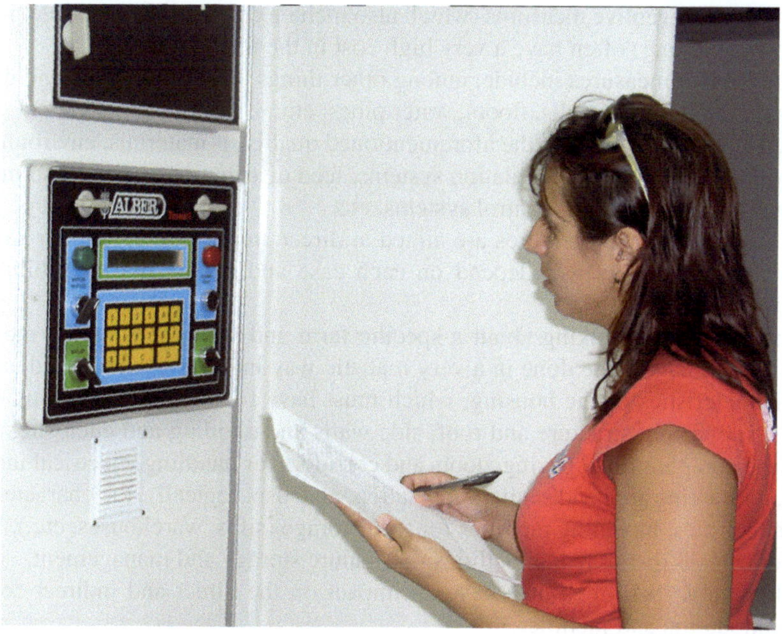

Fig. 7.3 Farms should be mechanised as much as possible, but this should be the result of projects tailored to well-planned production objectives

In practice, we will never be able to have the perfect housing and/or facilities on a livestock farm. But it should not be forgotten that, in livestock farming (as in other economic sectors), perfect is the enemy of good.

The following measures should be considered when designing our farm:

- Preventive or pre-analysis measures.
- Passive measures.
- Active measures.

Preventive measures include, for example:

(a) Analysis of the surface and characteristics of the available land
(b) The availability of water and electricity, which will condition the viability of our project
(c) Geographical location with respect to other farms, urban centres, transportation routes, industries, sources of raw materials, nearby noise, etc.
(d) The climate of the area, which will condition the characteristics of the accommodation
(e) Accessibility

Errors in preventive measures (which also include calculation errors, e.g. in ventilation or lighting) often have a very high cost in the medium term.

The passive measures include, among other things, the degree, type and quality of insulation in roofs, walls, floors, water pipes, etc.

Active measures include the aforementioned quality of materials, environmental control systems and their regulation systems, feed distribution systems and manure removal systems, lighting control systems, etc.

Most of the active measures are aimed at direct and/or indirect labour savings. Their appropriateness will depend on each case and on the real possibilities of amortisation.

Logically, when talking about a specific farm and when considering the measures, it will have to be done in a very realistic way in order to define and analyse the characteristics of the housing, which must have foundations, including earthworks, supporting structure and roof, side walls and dividing and enclosing walls, drainage and plumbing, paving, doors and corridors for handling, electrical installations, quarantine area and sanitary insulation, and also, logically, the characteristics of the facilities, including facilities for feed storage (silos, warehouses, etc.), facilities for carcass disposal and facilities for manure storage and management.

Ultimately, these measures have an impact on the direct and indirect costs of accommodation and facilities.

Within the framework of the livestock housing and facilities recommended by Prof. Buxade (Castelló 1996; Buxade 1998), and complementing what has been explained so far, the following factors or construction principles should be taken into account:

- Design work units that can be executed by direct labour present in the economic environment. Obviously, we are referring to general units, not those that may require specialisation.
- Seek the use of "non-specialised" material, which is larger in volume and, in general, may deteriorate more frequently and which comes from the geographical and economic environment of the farm.
- Specialised equipment should always be purchased from companies whose technical and commercial solvency (after-sales service) is guaranteed.
- Use, as far as possible, prefabricated construction elements, as the most complex construction process is carried out at source. This gives a greater guarantee of uniformity and quality, and on the other hand, the assembly of the building is simpler and quicker.
- Use, as far as possible, uniform and standardised construction elements, as this will greatly facilitate their implementation. The more identical or similar elements are used, the faster the work can go.
- Use construction elements and equipment that are easy to transport and install. As a general rule, materials that are easy to assemble are easy to replace.

In livestock farming, where we always work with living beings, more and more demands are being placed on environmental conditions; poorly made investments in housing and facilities are not only impossible to amortise, but in the short term, they can also affect the viability of the farm.

Breeder Housing

The guinea fowl is an animal that requires the use of artificial insemination techniques for reproduction. This is because they are monogamous animals, i.e. a male will only breed with the same female. Obviously, this implies that the reproduction rate will be low and that we must have as many males as females.

In order to optimise the reproductive indexes, the females and males are kept separately, so that semen can first be extracted from the males, after which the females can be inseminated. In order to carry out these practices, it is advisable to arrange the animals in batteries of cages.

The evolution of the batteries over the years has been towards an increase in the density of birds per building and towards greater mechanisation and greater automation. The types described by Professor Buxade (Castelló 1996; Buxade 1998) are as follows:

- California-type battery: block of different levels dispersed in a staircase in such a way as to allow the free fall of the excrement to the ground or pit.
- Flat-deck battery: a set of cages on the same level, with the corresponding equipment shared by several cages.
- Compact-type battery: modification of the California-type cage, reducing its width and increasing its height, which makes it necessary to carry out certain mechanisations to evacuate the droppings between floors.
- Inverted-type battery: the space of the trough is favoured over a greater investment by reducing the number of birds per linear metre of battery and the floor.
- Belt-type battery: modification of the Compact type in that the verticality is absolute and the collection of droppings is carried out by means of an endless belt located under each floor. Multiple heights can be designed, including an intermediate inspection corridor.

To design the breeder room, it is very convenient to have decided what type of battery we want to install, height, length, number of rows and height of the aisles. Through knowing the type of battery, we can automatically know the height. The length and number of rows are an economic compromise between:

- Longer length, lower battery costs.
- More rows mean more building width and therefore more economy in the building.
- Number of birds to be housed, which will be the link between length and number of rows.

This means that the cost per bird in a house of 100 m in length and 16 m in width is cheaper than a house of 75 m in length and 12 m in width.

Depending on the type of droppings collection in buildings for laying hens, they can be classified into three types:

- Deep pit house: compact, multi-storey California-type batteries are used, in which the excreta fall to a floor below the floor of the house aisle. The advantage of this system is the large waste storage capacity. The waste produced is dry, due to the residence time and the fermentation produced, 15–20% humidity. The disadvantages are the cost of implementation, air quality versus ventilation and insects breeding in the rubbish store.
- Warehouses for waste batteries by belt: this is an open-plan warehouse with a pit to house the transversal belt for the collection of excrement and exit to the outside. The greatest advantage of this system is the quality of the environment, especially in systems of air blowing into the battery for drying the waste. The main disadvantage is the cost of the battery and the air blowing systems.
- Ships for liquid waste: in compact and California-type batteries, the waste falls freely into pits, between 30 and 80 cm deep, at the bottom of which a shovel drags the waste towards one of the outer sides of the ship for its subsequent evacuation to a pit, from where it is stored. This type of waste can only be handled by vacuum tankers and with the help of mixers. Its humidity percentage ranges from 80 to 90% depending on whether more or less water is added. The advantage of the system is its economy, and the disadvantage is that it is very difficult to remove the waste from the farm.

Breeding Accommodation

The guinea fowl is a very skittish animal, and, although in its first breeding phase it does not require large spaces, it will be demanding in terms of available surface area per animal in the rearing phase.

The rearing phase is the most critical phase for the chick, because it is the phase in which the highest mortality rate occurs.

It is recommended that the animals are housed in a closed house, in fences that avoid angles, on a floor with clean straw or shavings. This is the best adaptation for this stage and allows the best control over the animals. Groups should not be excessively large, with a recommended density of 40 to 50 chicks/m2 in the first week—fences should be used inside the house to concentrate the chicks—and 15 to 20 chicks/m2 between the second and fourth week.

In the rooms, the floor should be made of concrete, without corners where possible. Each room shall have a window or double door to the communication corridor for control without the need to enter the room (Fig. 7.4). Thermal insulation of ceilings and external walls is recommended to allow environmental control and energy saving.

Fig. 7.4 Farms should be mechanised as much as possible, but this should be the result of projects tailored to well-planned production objectives

Fences may have a height of 40–50 cm. They should be arranged in such numbers that combined they contain a number of animals equal to the final composition in each of the rearing and finishing modules.

Each room shall have sufficient window openings for ventilation, with mosquito netting. The rearing rooms shall also have a hatch to allow passage of the chicks to the rearing pens.

Rearing Accommodation

In the case in question, the guinea fowl are moved to free-range, open-air runs from the 6th to 7th week of life, where they will remain until slaughter age (13–14th week). The surface area should be sufficient for each guinea fowl to have a little more than 1 m/bird and to protect the birds from temperatures and rain, a shelter with metal sheeting on the roof and the side where the coldest winds come from will be sufficient (Fig. 7.5).

Where possible, native herbaceous vegetation and trees (Fig. 7.6) should be provided for shading in summer.

Fig. 7.5 It is recommended that rearing rooms have double doors so that the birds can be inspected without entering the room

Fig. 7.6 In extensive farming systems, shelters in exercise yards consist of metal sheets to cover the roof and the side exposed to the coldest winds

Other Accommodation

This section includes the facilities necessary for the handling of eggs, from collection to hatching of the chicks. These include reception, selection and cleaning, disinfection chamber, storage, incubation and hatching.

All these rooms should be located in the same building, but with separate entrances; there should be actual physical separation of each room through a connecting corridor.

Circulation, or direction, should be from clean to dirty areas.

In the same building, there are usually rooms for offices or services, feed storage, tools, etc. meaning that everything is located in the same building, in such a way that is practical.

Orientation of the Facilities

Buildings intended for breeding birds, at least those with windows, should be oriented along their longitudinal axis from east to west, in order to allow maximum sunlight on the south façade, while the north façade is always shaded and cooler. However, in buildings with parks, their orientation may be more arbitrary, due to the fact that birds spend most of their lives outside the buildings, even at night.

The north-south orientation allows the rearing rooms and parks on the east side in the morning and those on the west side in the afternoon to be exposed to the sun for approximately the same amount of time. However, depending on various factors, other orientations can be used, such as placing the longitudinal axis of the building in a NE-SW or NW-SE direction.

The choice of location for a livestock farm is subject to a series of conditioning factors: regulations, infrastructures, waste disposal, etc. In general, healthy land is sought, protected from strong winds, but aerated, dry and well drained, avoiding excessively close obstacles that could interfere with ventilation, hills very exposed to the wind that could produce an excess of air intake and boxed-in, insufficiently ventilated, humid and very hot places.

The orientation of the building is particularly important when ventilation is natural or static. In principle, it is advisable to arrange it perpendicular to the prevailing winds, i.e. if the prevailing winds come from the N and S, the appropriate longitudinal orientation of the building would be E-W with an open front to the S. In this way, good natural ventilation of the building would be achieved, avoiding the appearance of turbulence.

Professor Buxade (Castelló 1996; Buxade 1998) recommends that unless the winds prevail, in warm areas, the longitudinal axis should be arranged in an E-W direction to achieve the following:

- The area exposed to the west should be as little as possible, so that there is no strong sunstroke in summer.
- On the south-facing facade, the winter sun, which rises little at the zenith, will penetrate well through the windows (drying out the bedding), but in summer, the eaves of the building will act as a sunshade and thus protect the livestock from direct sunlight.
- Natural ventilation is achieved by having a permanently warm and a permanently cold facade.

To facilitate the management of natural ventilation, the use of natural windbreaks can be of interest, as they reduce energy losses due to ventilation, as the structure is less exposed to winds, as well as shading the accommodation in summer.

In halls with natural ventilation systems, we recommend planting vegetation with an air permeability of 50%, thus providing wind protection over a distance of approximately 20 times the height of the hall.

Thermal Insulation

The main function of livestock buildings is to protect the animals housed in them from the weather outside. To this end, they must be made of materials with hygrothermal characteristics such that comfortable environmental conditions can be maintained inside. A well-insulated building will mainly contribute to reduce heat losses in cold weather, reduce heat gains in hot weather and optimise the performance of the air-conditioning installations, heating, cooling and ventilation, allowing them to operate at a moderate rate in such a way as to reduce the energy expenditure generated and extend the working life of the equipment.

On the contrary, poorly insulated livestock buildings with poor air-conditioning systems favour the presence of high concentrations of different gases produced by livestock and their faeces, providing an aggressive environment for the structural elements of the building and a harmful living environment for the farmer and the livestock.

Water vapour is one of the most frequent gases in livestock farms, which can be detected visually when condensation appears on walls and ceilings. Condensation occurs on the coldest surfaces and is therefore a marker of thermal bridges and poorly insulated elements.

The appearance of condensation contributes to the degradation of the materials that form part of the building, mainly concrete and metals, which in the long term can lead to the ruin of the construction. This phenomenon can be attenuated, or even avoided, by providing these constructions with good insulation.

The insulating capacity of a material is characterised by its thermal conductivity (λ), defined as "the amount of heat that passes through a sheet of material with a surface area of 1 m^2 and a thickness of 1m in 1 hour, when a temperature difference of 1°C is established between its faces".

In other words, the lower the λ, the greater the insulating capacity of the material, and materials whose λ is equal to or less than 0.15 Kcal/h m °C are considered thermal insulators.

Thermal insulators (Table 7.1) are materials of different nature and composition, which generally contain a significant amount of gas enclosed inside their cells or are formed by a network of fibres.

Depending on their nature, they can be vegetable (cork, wood agglomerate, compressed straw, etc.), mineral (glass wool, rock wool, expanded clays, perlites, etc.) and synthetic (polyurethanes and polystyrenes in sheets and foam, etc.).

Table 7.1 Characteristics of some materials used as thermal insulators

Materials	Density (Kg./m³)	λ (Kcal./h m °C)
Cellular concrete	400	0.094
Expanding clay	300	0.073
Expanding perlite	130	0.040
Cork agglomerate	110	0.034
Glass wool	10–90	0.030
Rock wool	30–150	0.034
Expanded polystyrene type I	10	0.039
Expanded polystyrene type III	15	0.033
Expanded polystyrene type IV	20	0.031
Expanded polystyrene type V	25	0.030
Expanded polystyrene type VI	30	0.029
Extruded polystyrene	30–35	0.024
Polyurethane-shaped plate	30–35	0.020
Polyurethane foam in situ	30–35	0.020

In order to evaluate the insulation provided by a wall, it is necessary to calculate the thermal transmission coefficient "K" defined as "the amount of heat that passes through an enclosure with a surface area of 1 m^2 and thickness of 1 m in 1 hour, when a temperature difference of 1 °C is established between its faces".

Ventilation

Ventilation is one of the critical factors on livestock farms. Its functions include the permanent renewal of air, providing oxygen; the evacuation of noxious gases, odours and dust; the control of ambient temperature and humidity; and the dilution of pathogens (Kim and Lee 2022; Boutera et al. 2024).

The principle of ventilation is simple: fresh air from outside enters the housing; mixes with the air inside; takes in heat, humidity and suspended matter; and leaves the room, thanks to the various forces that cause the air to move (Fig. 7.7).

It is precisely the elements that produce these forces that serve to differentiate the two main ventilation systems: dynamic or forced ventilation and static or natural ventilation.

Natural Ventilation

In this system, air movement is produced by pressure gradients derived from natural phenomena such as temperature differences or the action of the wind between one area of the housing and another and between the outside and the inside of the

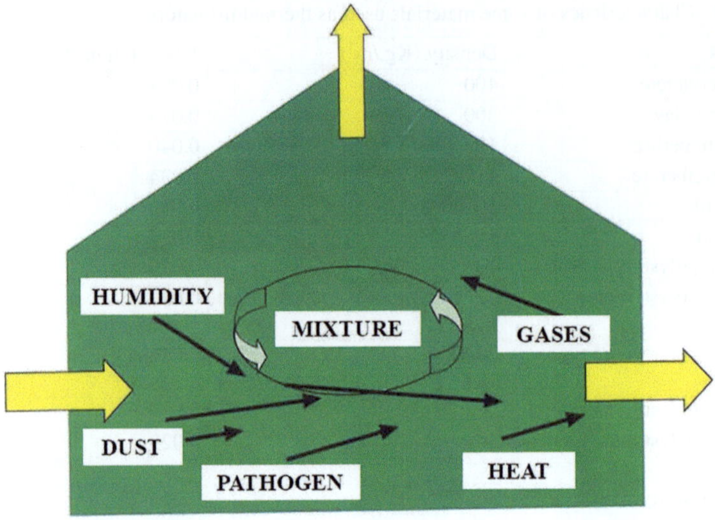

Fig. 7.7 Diagram of ventilation in a building

housing and which depend on atmospheric conditions, the design and orientation of the building, the existence of obstacles in the vicinity of the building, etc.

It can be deduced that natural ventilation has numerous constraints and limitations, and its performance will depend, among other factors, on the placement and design of the building openings through which air enters and exits, the temperature difference between indoors and outdoors, the slope of the roof, the orientation of the building with respect to the prevailing winds and their velocity, the height of the building, the indoor air velocity and the exposure to these air currents.

Natural or static ventilation is based on the formation of natural air currents, which are generated by two basic principles: the difference in density and the difference in pressure.

One of the causes of natural air movement is the so-called chimney effect (Fig. 7.8) or, to use more technical terms, thermal buoyancy. When the air enters the housing, it is heated by the heat dissipated by the animals and the fermentation of the droppings: as its specific weight decreases, it rises and escapes to the outside through the openings provided for this purpose, either chimneys or, more usually, a continuous opening in the ridge of the roof. The air, on leaving, creates a slight depression in the interior that causes air to enter from the outside as the pressure at this point is slightly higher than the pressure inside.

The "wind effect" (Fig. 7.9) is the second of the forces acting on natural ventilation, so that its effect is greater than the stack effect the higher its speed, and the more similar the external and internal temperatures are (as is the case in summer), a situation in which the stack effect is of little importance.

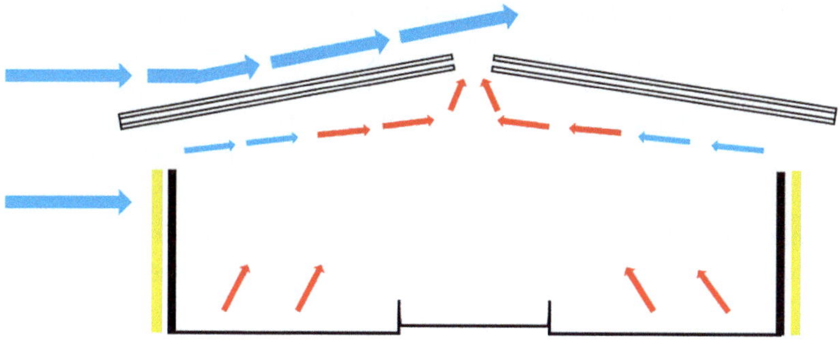

Fig. 7.8 Diagram of the chimney effect in natural ventilation

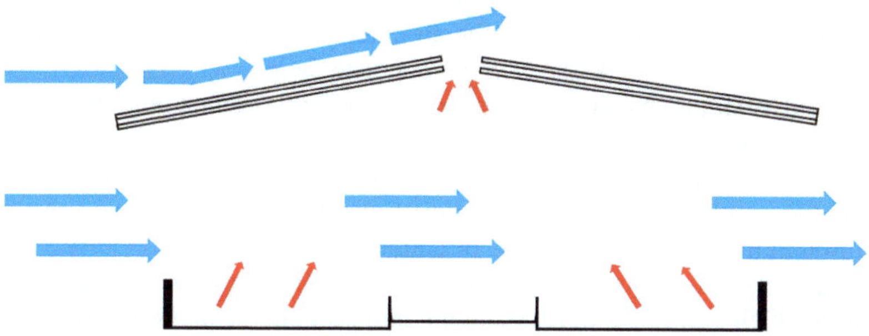

Fig. 7.9 Diagram of natural ventilation due to the wind effect

Dynamic Ventilation

In this system, air is blown in or out of the building by fans with a certain flow rate and, most of the time, with operation controlled by more or less sophisticated control systems.

There are three types of dynamic (also called forced or mechanical) ventilation that differ in the relative pressure of the air inside the enclosure in relation to the external atmospheric pressure:

- Extract ventilation: the fans extract air from the housing, creating a slight depression with respect to the outside, which means that fresh air enters through the inlets provided for this purpose.
- Injection ventilation: the fans inject fresh air into the housing, causing a certain overpressure with respect to the outside, overpressure that expels the inside air through the existing outlets.
- Balanced ventilation: some fans inject fresh air and others extract fresh air from inside the housing at a similar rate, so that the internal pressure is equal to atmospheric pressure.

Table 7.2 Dynamic ventilation options depending on the width of the building

Width of the ship	Air circulation in the ship	Ticket locations air	Location of exits air
<8 m	Transversal	A facade	Opposite facade
8–17 m	Transversal	Both facades	A facade
>17 m	Longitudinal	One end of the ship	Opposite end

Air injection requires perforated ducts to ensure adequate air distribution in the space occupied by the animals. Extraction ventilation is the most common.

Design of Ventilation Systems

When a building is subdivided into separate rooms arranged like a train carriage, the reasonable solution is one in which the air enters each room from the general corridor and exits to the outside through the opposite façade.

In the other situations, different options are possible, depending on the width of the space in question (Table 7.2).

For effective ventilation, the air inlets should be located at the top of the enclosures and the outlets relatively low, although the air can be released to the outside even through the roof (e.g. by means of chimneys).

The choice of fans must take into account their effective flow rate under the conditions in which they are to operate, taking into account the pressure losses suffered by the air.

If the design of the air inlets and outlets does not allow the accommodation to be temporarily ventilated by static ventilation in the event of a power failure, emergency equipment (windows or chimneys) should be installed to ensure some air renewal until the exhaust fans are restarted.

Type of Fans

To ensure that the actual ventilation flow rate is adjusted to the changing air renewal requirements (which can vary over a wide range depending on the specific circumstances at any given time), the use of fans with adjustable flow rates is advisable. The use of fixed-flow fans would lead to abrupt jumps in the ventilation flow rate, resulting in significant fluctuations in indoor climate conditions, which would rarely correspond to the desired conditions.

Air Inlet Velocity

The speed at which air enters the housing conditions its trajectories inside, affecting the efficiency of ventilation and the air speed and temperature at the height of the animals. When the air inlet into the housing is horizontal (the usual situation), high inlet velocities should be sought: 4–5 m/s (and even higher in very wide houses).

Air Inlet Openings

Given the wide range of variation in ventilation flow rates in a given enclosure, and to ensure that the air inlet velocity is always acceptable, it is necessary that the openings through which the air enters the enclosure are of variable cross-section and regulated according to the flow rates at any given time.

It is also preferable that the total cross-section of the air inlets is spread over several small openings rather than a few large ones: ventilation will be more uniform, and the temperature and air velocity at animal height will be improved.

Air Speed at Animal Height

The air speed at animal altitude influences, on the one hand, the heat loss of the animals (which increases as the air speed increases) and, on the other hand, the incidence of respiratory pathologies.

Animal Welfare Requirements

We have to deal with the controversial topic of the welfare of hens. Welfare and stress can hardly be related, as the former refers to psychological aspects and the latter to physical aspects. On the contrary, it seems logical to associate welfare with production (Dobraszczyk 2024; Wilcox et al. 2024).

Breeders

In addition to the general requirements laid down in the relevant legislation, enriched cages shall meet the following requirements:

- The height of the cage other than above the useable area shall be at least 20 cm at any point, and the total area of the cage shall not be less than 2000 cm^2. The usable area is defined as an area of at least 30 cm wide, with a maximum slope of 14% and with a clear space of at least 45 cm in height. The nest surfaces are not part of the usable area.
- A nest shall be provided.
- They shall be provided with a litter for pecking and scavenging.
- They shall have suitable perches providing at least 15 cm space per hen.
- A feeding trough shall be provided which can be used without restriction. Its length shall be at least 12 cm multiplied by the number of hens in the cage.
- Each cage shall be provided with an appropriate drinker, taking into account, in particular, the size of the flock. In the case of connected drinkers, at least two nipples or two cups shall be within reach of each hen.

- To facilitate inspection, installation and removal of animals, rows of cages should be separated by aisles at least 90 cm wide, and there should be a space of at least 35 cm between the floor of the establishment and the cages in the rows below.
- The cages shall be equipped with appropriate claw clipping devices.

Baiting of Chickens

In addition to the corresponding legal regulations, the main requirements to be fulfilled are the following:

- The maximum stocking density on the holding or in a house on the holding shall not exceed 30 kg/m^2 or, failing that, 15 birds/m^2, where stocking density is defined as the total live weight or the number of birds per square metre of usable area (space with litter accessible to the chickens at all times) present at any one time in a house.
- Training: keepers (persons responsible for or in charge of the care of chickens) shall have appropriate training, through the completion of courses approved by the Administration and accredited by certificates recognised by the Competent Authority. In addition, the owner or keeper shall provide written instructions and guidance on the relevant animal welfare requirements, including, where appropriate, those relating to on-farm slaughter methods, to persons employed or engaged by them to care for chickens or to catch and load them.
- Requirements for rearing chickens:

  1. Water troughs. Water troughs shall be located and maintained in such a way that water spillage is minimised.
  2. Feeding. Feed shall be continuously available or fed by meal and may not be withdrawn more than 12 hours before the scheduled slaughter time.
  3. Litter. All chickens shall have permanent access to dry, friable surface litter.
  4. Ventilation and heating. Sufficient ventilation should be provided to prevent overheating and, where appropriate, combined with heating systems to remove excessive humidity.
  5. Noise. Noise levels should be kept as low as possible. Fans, trough systems and other equipment shall be constructed, erected, maintained and operated so as to produce as little noise as possible.
  6. Lighting. All housing shall have lighting with a minimum intensity of 20 lux during daylight periods, measured at bird's eye level, illuminating at least 80% of the usable area. Where necessary, a temporary reduction of the lighting level may be authorised on veterinary advice. Within 7 days from the time when the chickens are placed in their housing until 3 days before the intended time of slaughter, lighting shall follow a 24-hour rhythm and include periods of darkness lasting at least 6 hours in total, with a minimum period of uninterrupted darkness of 4 hours, excluding periods of semi-darkness.

7. Surveillance. All chickens on the holding shall be inspected at least twice daily. Particular attention shall be paid to signs indicating a decrease in the welfare or health status of the animals. Chickens with serious injuries or with obvious signs of health disorders likely to cause pain, such as those with gait difficulties, severe ascites or major malformations, shall receive appropriate treatment or be culled immediately. A veterinarian shall be notified whenever necessary.
8. Cleaning. Those parts of the facilities, equipment or utensils which come into contact with the chickens shall be thoroughly cleaned and disinfected each time a complete depopulation is carried out before a new flock is introduced into the house. After final emptying of a poultry house, all litter shall be removed, and clean litter shall be provided.
9. Record keeping. The following particulars shall be entered by the owner or keeper in the holding register, or in a specific register, for each house on a holding:

    (a) Date of arrival of the animals and number of chicks introduced
    (b) Usable area
    (c) Breed or breed of chickens if known
    (d) For each control, the number of birds found dead indicating the causes, if known, as well as the number of birds culled for this cause
    (e) Date of departure of the animals and number of chickens leaving
    (f) The number of chickens remaining in the flock after the removal of chickens for sale or slaughter. These data must be kept for a period of at least three years so that they can be presented to the competent authority when carrying out an inspection or when otherwise requested.

10. Surgical interventions. All surgical interventions for non-therapeutic or non-diagnostic reasons which may result in injury or loss of a sensitive part of the body or alteration of bone structure are prohibited. However, the competent authority may authorise the beak trimming of birds after all other measures to prevent feather pecking and cannibalism have been exhausted. In such cases, the operation shall only be carried out after consultation with and on the advice of a veterinarian and shall only be carried out by qualified personnel and only on chicks less than 10 days old. The competent authority may authorise the castration of chicks, which may only be carried out under veterinary supervision and by specifically trained personnel.

Competing Interests The author has no conflicts of interest to declare that are relevant to the content of this chapter.

References

Adams N, Sans A, Trier Kreutzfeldt KE, Arias Escobar MA, Oudshoorn FW, Bolduc N, Aubert PM, Smith LG (2025) Assessing the impacts of EU agricultural policies on the sustainability of the livestock sector: a review of the recent literature. Agric Hum Values 42(1):193–212

Bist RB, Bist K, Poudel S, Subedi D, Yang X, Paneru B, Mani S, Wang D, Chai L (2024) Sustainable poultry farming practices: a critical review of current strategies and future prospects. Poult Sci 103:104295

Boutera Y, Boultif N, Moummi N, Arıcı M, Saleh MS, Rouag A, Kethiri MA, Beldjani C (2024) Evaluation of the earth-air heat exchanger's performance in improving the indoor conditions of an industrial poultry house using computational fluid dynamics verified with field tests. J Clean Prod 434:140218

Buxade C (1998) Zootechnics. Bases of animal production. Monograph II. Alojamientos e instalaciones II. Editorial Mundi-Prensa, Madrid

Castelló JA (1996) The exploitation of Guinea fowl. In: Zootecnia. Bases de Producción Animal. Volume X. Editorial Mundi-Prensa, Madrid

Dobraszczyk P (2024) Beyond domesticities: Posthuman architectures for animals we farm. Archit Des 94(1):76–83

Dupas MC, Pinotti F, Joshi C, Joshi M, Thanapongtharm W, Dhingra M, Blake D, Tomle F, Gilber M, Fournié G (2024) Spatial distribution of poultry farms using point pattern modelling: a method to address livestock environmental impacts and disease transmission risks. PLoS Comput Biol 20(10):e1011980

Kim SJ, Lee MH (2022) Design and implementation of a malfunction detection system for livestock ventilation devices in smart poultry farms. Agriculture 12(12):2150

Nielsen SS, Kemunto NP, Muloi DM, Bojesen AM, Knight-Jones T, Ayebare D, Dione M, Bayi EL, Poulsen LL, Moodley A (2025) Flexible assessment of biosecurity in small-and medium scale poultry farms in low and middle income countries. Acta Vet Scand 67(1):9

Wilcox CH, Sandilands V, Mayasari N, Asmara IY, Anang A (2024) A literature review of broiler chicken welfare, husbandry, and assessment. Worlds Poult Sci J 80(1):3–32

Zamani O, Chibanda C, Boimah M, Asante-Addo C (2025) Aligning policy for success in developing countries: evidence from the poultry sector of Ghana. Agric Food Econ 13(1):1–19

Meat Quality

8

Susana García-Torres, María Montaña López Parra, Carmen Sanabria Tienza, David Tejerina Barrado, and María Jesús Martín Mateos

Abstract

Meat quality is determined based on its nutritional value, taste and food safety, which are assessed through physical, chemical, sensory and microbiological tests. Physical and chemical analyses of guinea fowl meat are aimed at measuring colour, pH, meat texture, intramuscular fat content, protein and ash content. The sensory quality of the meat is assessed by a panel of trained tasters who evaluate the intensity of the poultry smell, other odours present, tenderness, juiciness, oiliness, fibrousness, intensity of the poultry flavour and overall appreciation. Experiments carried out to determine the maximum storage time of the meat without loss of food safety or organoleptic quality show that, through pretreatment with lactic acid followed by ozonation of the carcasses, followed by vacuum packaging and refrigerated storage, the treated samples remain stable for 10 days without any loss of quality. These findings set a new benchmark for the techniques under development, which also allow for improved traceability throughout the process.

Keywords

Meat quality · Guinea fowl · Preservation treatments · Organoleptic quality

S. García-Torres (✉) · M. M. L. Parra · D. T. Barrado · M. J. M. Mateos
Meat Quality Area at CICYTEX (Scientific and Technological Research Centre of Extremadura), Badajoz, Spain
e-mail: susana.garciat@juntaex.es; montana.lopez@juntaex.es; david.tejerina@juntaex.es; mariajesus.martinmat@juntaex.es

C. S. Tienza
CICYTEX (Scientific and Technological Research Centre of Extremadura), Badajoz, Spain

J. del Moral de la Vega, J. del Moral Martínez (eds.), *The Guinea Fowl (Numida meleagris)*, https://doi.org/10.1007/978-3-032-11654-3_8

Quality is a very complex term that scores different aspects of the product depending on the perspective from which it is assessed. In the case of meat, we could say that meat quality refers to three fundamental aspects (FAO and WHO 2024):

- Nutritional—the set of chemical compounds responsible for the nutritional characteristics of a product
- Sensory—the set of physical attributes perceived by the senses that are responsible for the overall appearance of a product
- Microbiological—linked to food safety, which is regulated, with limits that determine whether a product is fit or unfit for consumption

Overall, understanding the quality of a product means, from a business point of view, the possibility of enhancing its best attributes and improving the aspects of a lower quality.

Guinea fowl meat has enjoyed recognised prestige since Roman times. Its qualities have made it a highly acceptable culinary product, occupying a privileged place in French cuisine and in haute cuisine (Cepero et al. 2005; Bernacki et al. 2013).

Guinea fowl meat is said to be reminiscent of pheasant meat and "has an unmistakable but more delicate flavour than game", and from a gastronomic point of view, it lends itself to the development of a wide variety of recipes, with preparation techniques ranging from the simplest to the most sophisticated (Fig. 8.1) (Hoffman and Tlhong 2012). It is therefore in great demand, especially by consumers looking for superior quality poultry meat.

Fig. 8.1 Guinea fowl recipe by David Chapela, chef at Hotel Zurbarán, Badajoz (Extremadura, Spain) (mushrooms stuffed with guinea fowl and foie gras with walnut vinaigrette)

The result is a different product to the one known classically and usually produced in France in intensive production systems (Cepero et al. 2005).

The carcass of these birds is quite different from that of any other poultry we are used to eating. The skin is darker in colour, and the average weight is 1.5–1.7 kg.

Meat quality studies are generally carried out by physico-chemical analysis of the meat, which allows the product to be characterised objectively.

Physico-Chemical Quality of Guinea Fowl Meat

The quality of meat products depends on multiple factors and is affected by each and every one of the processes involved in obtaining them, from animal production to handling by the consumer.

In general, guinea fowl offers all the elements of a healthy meat, low in calories, low in cholesterol and very rich in protein. The thigh meat has approximately 2.7% fat, and the breast meat has 1.1% fat, which makes it the lowest fat poultry. It also has a high percentage of free fatty acids and a minimal amount of cholesterol. The protein content is higher than that of chicken, with high amounts of vitamins B1, B12 and E, as well as a high percentage of minerals, such as magnesium, calcium and iron (Musundire et al. 2017; López-Pedrouso et al. 2019).

Among the different studies that have been carried out on guinea fowl, it is worth highlighting an exhaustive assessment of the quality of the meat produced in southwestern Spain (Extremadura), where birds were reared in two very different production systems—intensive and extensive—and the quality of the meat was assessed throughout the four seasons of the year, specifically when guinea fowl were reared extensively (García Torres et al. 2009). Different studies show the importance of production location and quality of guinea fowl meat (Szalay et al. 2016; Shoyombo et al. 2021; Erensoy 2025).

The physico-chemical quality of the meat is defined by the classic parameters in the quality protocols for fresh meat, which depend on the piece of meat being analysed; in this case, the two main cuts are the breast and the thigh.

Meat Colour

One of the factors that play a decisive role in the assessment of meat quality is colour. It is a primary attribute in the acceptability of the meat by consumers, as well as in the purchase of the product, as consumers relate it directly to the freshness of the meat and the sensory properties of the meat. Poultry meat is the only meat sold with or without skin. We are used to expecting poultry meat to have a pale pink colour on the breast and a deeper red colour on the thigh (Barbut and Leishman 2022).

In general, in poultry, the breast and wings are lighter in colour than the thighs and legs, a consequence of the amount of exercise performed and the concentration of myoglobin in the piece, as well as the type of muscle fibre. The main differences between the breast (white meat) and the thigh (dark meat) is that the breast has less fat and less connective tissue and takes less time to cook, unlike thigh meat.

The colour of the meat depends on several factors. The colour is determined by the amount of myoglobin and haem pigments in the muscle, which is the most important intrinsic factor, and is related to breed, sex, production system, age, slaughter weight, etc. Colour is also indirectly influenced by pre- and post-slaughter conditions, such as stress, temperature and humidity in the chambers. These factors alter the rate of pH decline in the meat, which is directly related to multiple quality parameters. The meat preservation process, storage conditions and storage time can also alter meat colour.

Colour measurement is carried out in several ways. Colour can be assessed subjectively (visual colour perception) and objectively (physical or instrumental and chemical) by using a colourimeter (Fig. 8.2), which physically breaks down colour into three parameters L^* (lightness), a^* (red index) and b^* (yellow index).

Colour can also be assessed chemically by measuring the amount of haem pigments in the meat (CIE 2025). The colour of the meat depends mainly on the state of myoglobin. Myoglobin is a small globular protein with a haem group containing an iron atom. It is present in the muscles of vertebrates and has the function of transporting oxygen to the muscles. The chemical form of myoglobin varies depending on whether or not it has an oxygen molecule attached: oxymyoglobin (brighter red colour) or metamyoglobin (browner colour). Its determination is based on the absorption of light in the visible spectrum using spectrophotometry.

The (subjective) perception of the colour of a guinea fowl carcass by an observer consists, firstly, in assessing the colour of the skin. In contrast to the poultry we usually cook, the skin of guinea fowl carcasses is dark (Fig. 8.3), with a thinner skin,

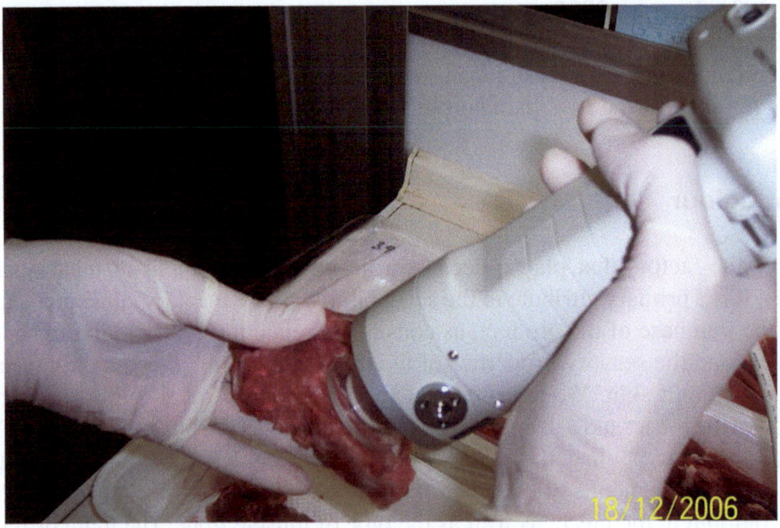

Fig. 8.2 Colourimeter used to determine the colour of the meat

Fig. 8.3 The skin of guinea fowl carcasses is dark in colour

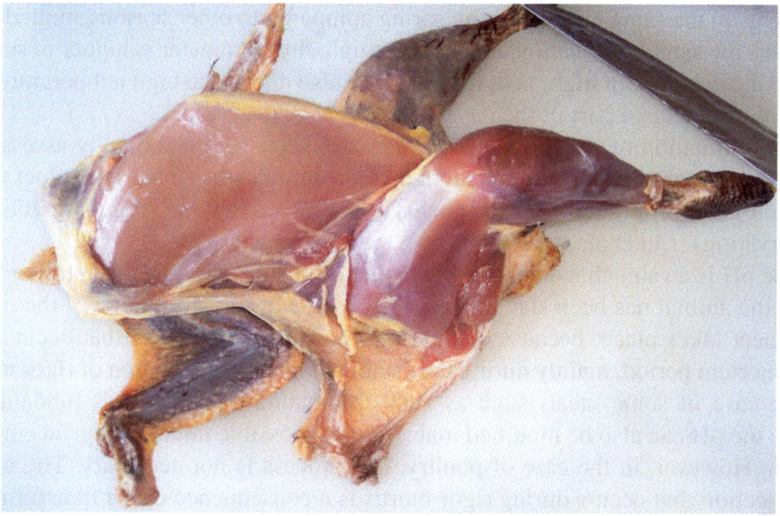

Fig. 8.4 The colour of guinea fowl meat is red

and the colour of the musculature can be seen through it, influencing the assessment of the skin colour.

As for the colour of the meat, the appearance is red (Fig. 8.4), much darker than usual for poultry, both in the breast and in the thigh.

In studies carried out in south-western Spain (Extremadura), to analyse skin colour using a colourimeter, no significant differences were found when comparing guinea fowl produced in different production systems (extensive vs. intensive), nor in the meat of any of the pieces studied (breast and thigh). The animals raised in the extensive system showed differences according to the season of the year. In this case, small differences in skin and meat colour were observed, indicating that breast skin in spring had a higher yellow index (b^*) than in other seasons, while thigh skin showed a higher red index (a^*) in the same season. Likewise, the colour of the breast meat showed a higher red index in spring, but it was the colour of the thigh that varied the most according to the season studied (López Parra et al. 2007) (Table 8.1).

The measurement of colour by chemical methods, in contrast to the physical measurement, determined that the quantity of pigments in the breast meat of guinea fowl varied according to the production system, being more abundant in guinea fowl raised in extensive systems than in intensive systems. These differences were also evident according to the different seasons of the year in both breast and thigh meat, with a greater abundance of haeminic pigments in breast meat during spring, while in thigh meat, there was a greater amount of pigments in summer (Fig. 8.5) (López Parra et al. 2005). These results correspond to the need for a greater oxygen supply in the *pectoralis superficialis* (breast) muscle in the case of guinea fowl raised in extensive systems, since they are flying fowl, and the open space, although fenced, allows them to do so. On the other hand, the increase in the amount of haeminic pigments in the same muscle during spring compared to other seasons studied probably has the same explanation, and for the thigh, this parameter is higher in summer due to the decrease in flight activity, probably also due to the high temperatures that occur during a large part of the day.

The estimation of pH is a very relevant parameter for meat quality assessment, since the evolution of pH after slaughter of the animals has a marked effect on the technological properties of meat quality (such as processing, transformation and preservation) (Liu et al. 2022).

The pH is a value that determines whether a substance is acidic, neutral or basic. Once the animal has been slaughtered, the process of transformation of the muscle into meat takes place, because of multiple biochemical changes that occur in the post-mortem period, mainly during the establishment and resolution of rigor mortis. In the case of some meats such as veal, the maturation process is fundamental, where the pH can also be modified mainly due to possible microbiological contamination. However, in the case of poultry, this process is not necessary. The muscle acidification that occurs during rigor mortis is a consequence of the transformation of muscle glycogen into lactic acid. Both the final pH value, which is the pH value in the muscle once rigor mortis is resolved, and which varies according to the species, and the rate at which it falls during the muscle-to-meat conversion affect the organoleptic and technological characteristics of the meat.

We could say that pH assessment estimates the existing level of lactic acid in the meat, which allows the assessment of post-mortem muscle glycolysis. The decrease in muscle glycogen rates varies, mainly due to the different factors that cause both physical and physiological stress in the animals: noise, transport, water and feed deprivation, handling, the disruption of established social structures or the regrouping animals from different origins to new structures, etc. Such is the importance of

Table 8.1 Colour of guinea fowl meat (*Numida meleagris*) measured instrumentally according to the seasons of the year

		Breast				Thigh			
		Autumn	Spring	Summer	Signif.	Autumn	Spring	Summer	Signif
Skin	L^*	54.6 ± 2	51.4 ± 2	58.3 ± 2.1	ns	50.7 ± 6	47.3 ± 7	48.9 ± 5	ns
	a^*	5.9 ± 0.6	4.2 ± 0.6	4.8 ± 0.6	ns	6.7 ± 2.4^b	9.6 ± 2.9^a	8.7 ± 1.9^a	*
	b^*	8.6 ± 2.3^b	13.8 ± 6.7^a	8.2 ± 6^b	*	3.1 ± 2	4.2 ± 2.4	2.4 ± 1.8	ns
Meat	L^*	52.3 ± 0.8	50.8 ± 0.8	51.9 ± 0.8	Ns	40.9 ± 2.5^b	44.02 ± 2.3^a	41.9 ± 2.8^b	*
	a^*	6.6 ± 0.4b	8.0 ± 0.4^a	6.7 ± 0.4^b	*	16.0 ± 3.6b	18.1 ± 1.9^b	13.4 ± 2.8^a	*
	b^*	7.8 ± 0.5	7.7 ± 0.5	9 ± 0.5	ns	5.5 ± 1.9^b	7.3 ± 1.5^a	6.9 ± 1.0a	*

L^*, lightness; a^*, red index; b^*, yellow index. Signif: statistical significance. n.s.: not significant. *: significance $p = 0.005$. Values with different letters (a, b) in the same row indicate significant differences ($p = 0.05$)

Fig. 8.5 The colour of guinea fowl carcasses raised in extensive farming varies according to the season

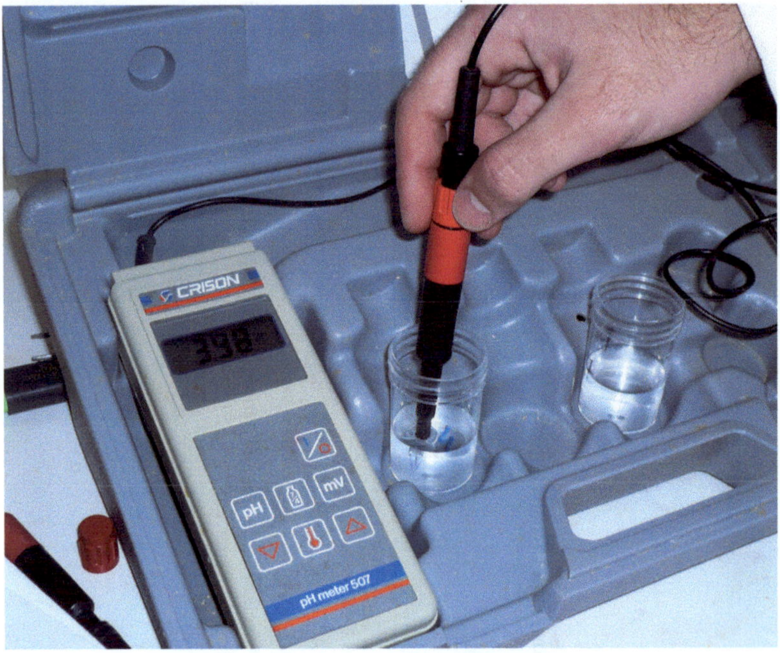

Fig. 8.6 Estimating pH is a very useful parameter for assessing meat quality

this parameter that a decrease in pH will affect the colour of the meat, the denaturation of the myofibrillar proteins, the decrease in the amount of water retained between them and, therefore, the amount of exudate in the meat and its texture.

The decline in pH in poultry is generally estimated between 6 and 8 h, when rigor mortis begins to break down, although each muscle has its own pattern. The pH studies with guinea fowl in south-western Spain (Extremadura) (Fig. 8.6) were carried out at 24 h post-mortem, when the pH was stable. They found that there was no difference in the final pH in the thigh of these poultry reared extensively versus those raised extensively, while the pH of the breast was higher in extensively versus intensively reared guinea fowl. In the comparison between seasons, the highest pH values were found, both in the breast and thigh, in the most extreme seasons in terms of climate, i.e. summer and winter (López Parra et al. 2005).

The final pH values, as has already been mentioned, have a direct impact on other very important quality parameters. Thus, a drop in pH results in protein denaturation, which leads to a reduction in the water contained between the myofibrillar proteins, resulting in the exudation of the meat. In general, the pH values obtained in guinea fowl are within the normal range for poultry meat.

The main component of meat is water, which is estimated at 75% on average, although it varies depending on factors such as the species of animal or the type of muscle considered; therefore, it is an interesting quality parameter to address. Although a high-water content is lost in meat handling processes, through the evaporation of water during cooling, handling and cutting, there is a reservoir of water that meat is capable of retaining and which determines many of its properties (Barbut 2024).

After the animal has been slaughtered, blood circulation and oxygen supply cease, the pH drops and protein denaturation begins, which in turn favours the exudation or release of water. This exudation will determine, among other things, the juiciness of the meat.

The ability of the meat to lose or retain water is therefore assessed. The breast part of the guinea fowl meat can retain more water than the thigh. In this case, the thigh muscles are more fibrous and metabolically different from the *pectoralis superficialis* (breast). On the other hand, it is physiological factors that determine the differences in this parameter, and not the production system.

Texture is a very important and highly valued property in meat quality, especially in other types of meat such as veal, where it is a fundamental parameter for consumers. It is a complex parameter to define, since it involves a multitude of characteristics or attributes of the meat. It is directly related to the myofibrillar structure, the connective tissue content and the cytoskeleton of the muscle, which later become components of the meat. It is highly variable and depends on factors such as species, breed, age and sex. Meat texture is characterised by the perception of tenderness and juiciness. It is determined by physical methods and measures the force needed to cut the meat (Fig. 8.7). This way of measuring meat tenderness is limited to the amount of force necessary to break the fibrillar structure, without considering other aspects such as juiciness or fat content that result in the consumer's perception of tenderness. In studies carried out on this variable, guinea fowl thigh meat was found to be less tender than breast meat. The results obtained showed that the production system did not affect the tenderness of the guinea fowl meat in any of the pieces studied. However, when studying the tenderness of the meat reared extensively across different seasons, it was determined that both the breast and thigh meat from the winter study were the most tender. Overall, the values obtained in the physical texture study indicated meat that was apparently firm yet tender (López Parra et al. 2005).

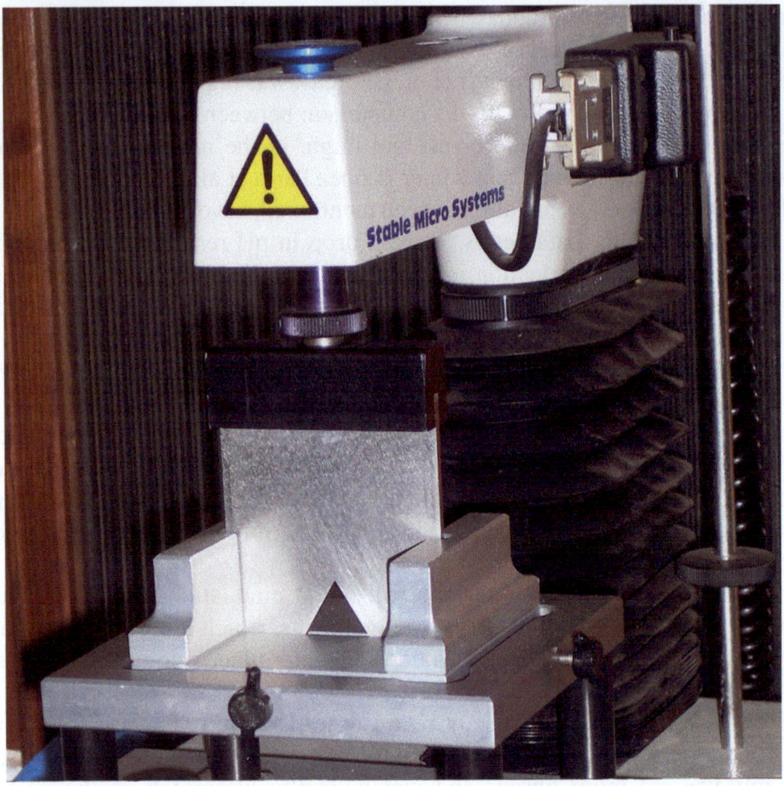

Fig. 8.7 Texture meter used to determine meat tenderness

Nutritional Quality of Guinea Fowl Meat

From a nutritional quality point of view, it is very important to know the intramuscular fat, protein and ash content. In general, poultry meat is a good source of protein, both in terms of quantity (20%–22%) and quality (Hailemariam et al. 2022). On average, 40% of the amino acids in meat are essential. Based on this profile, meat protein can be considered of high biological value. This is important because the human body needs the presence of all amino acids to synthesise proteins. The guinea fowl breast has a higher protein content than the thigh, and this relationship is reversed when the intramuscular fat content is assessed (Table 8.2). The amount of fat in poultry meat can vary significantly depending on the muscle from which the meat sample is taken.

The production system influences the nutritional quality of the guinea fowl meat, especially the breast of extensively reared guinea fowl, which is higher in protein and lower in fat than guinea fowl reared in the intensive production system (Tejerina et al. 2009).

Table 8.2 Nutritional content of *Numida meleagris* meat in extensive and intensive production systems (%)

			Breast		Thigh	
Content %	Breast	Thigh	Extensive	Intensive	Extensive	Intensive
Ashes	1.11 ± 0.2^a	1.07 ± 0.19^a	1.14 ± 0.18^a	1.09 ± 0.22^a	1.09 ± 0.14^a	1.05 ± 0.24^a
Proteins	24.7 ± 1.8^a	21.56 ± 1.85^b	25.55 ± 1.43^a	23.81 ± 1.72^b	21.78 ± 2.37^a	21.32 ± 1.1^a
Grease	1.01 ± 0.21^a	2.33 ± 0.98^b	0.92 ± 0.17^a	1.1 ± 0.21^b	2.24 ± 1.13^a	2.43 ± 0.82^a

$\bar{X}$ (mean) ± *SD* (standard deviation)

Values with the same letters (a, b, c) in the same row indicate non-significant differences ($p = 0.05$)

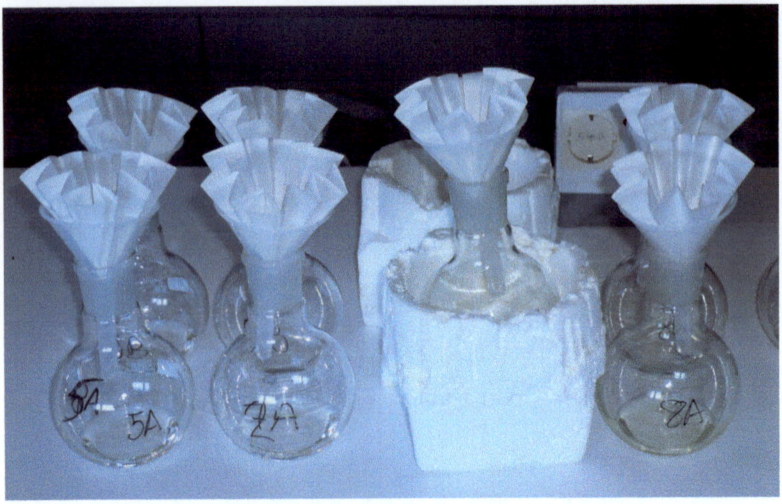

Fig. 8.8 Stage of the procedure used to determine the quality of fatty acids in meat

Table 8.3 Composition of major fatty acids in guinea fowl breast by system (%)

	Intensive	Extensive	Significance
C16:0	26.79 ± 3.77	26.91 ± 3.68	ns
C18:0	13.28 ± 2.21	13.25 ± 1.86	ns
C18:1 n9c	19.28 ± 1.76	19.30 ± 1.85	ns
C18:2	24.18 ± 1.91	23.77 ± 1.64	ns
C20:4 n6	7.59 ± 2.25	7.67 ± 2.02	ns
SFA	28.02 ± 4.07	28.12 ± 3.92	ns
PUFA	34.75 ± 3.35	34.74 ± 3.40	ns
MUFA	23.57 ± 2.50	23.55 ± 2.49	ns

$\overline{X}$ (mean) ± *SD* (standard deviation)
SFA total saturated fatty acids, *PUFA* total polyunsaturated fatty acids, *MUFA* total monounsaturated fatty acids, *ns* no significant difference

Due to the importance of fatty acids in human nutrition and health, it is important to study the quality of fatty acids in meat (Fig. 8.8).

The fatty acid profile of guinea fowl breast meat does not vary according to the production system (Table 8.3). There are only differences in the amount of some fatty acids, such as arachidic (C20:0) and eicosatrienoic (C20:3). Both show an inverse behaviour: C20:0 is higher in intensively produced meat, while C20:3 is higher in extensively produced meat. Poultry meat commonly found on the market has a higher percentage of saturated fatty acids, which are undesirable for health (Hailemariam et al. 2022), than the content in the meat of guinea fowl produced both extensively and intensively (García Torres et al. 2009).

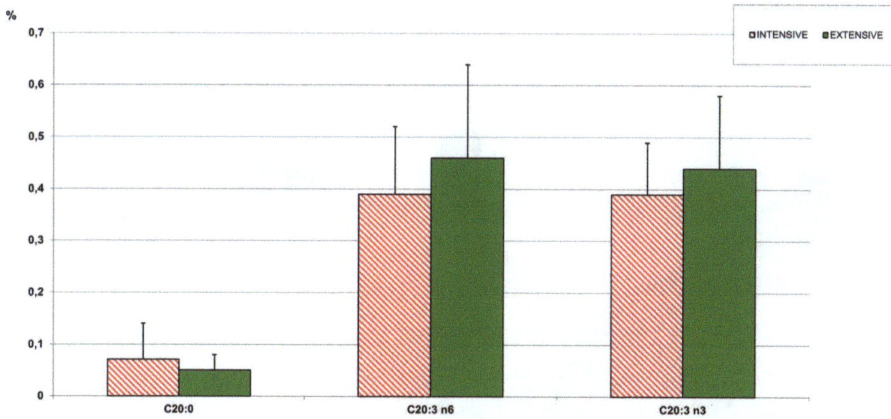

Fig. 8.9 Fatty acid content of the C20 group in guinea fowl breast meat in two production systems: extensive and intensive (*$p < 0.05$ statistically significant difference)

Moreover, the content of heart-healthy fatty acids (polyunsaturated) found in guinea fowl meat, regardless of the production system, is higher than that of broiler chicken meat.

Essential fatty acids are a group of fatty acids that cannot be produced by the body and must be ingested through food or supplements. In general, they regulate physiological processes related to cellular and physiological growth, brain structure and cell metabolism.

In guinea fowl meat, the content of essential fatty acids of group C20 (Fig. 8.9) must be highlighted, which are precursors of prostaglandins. These compounds play a very important physiological role in various functions such as renal and anti-inflammatory processes and in recent years have been recognised for their role against oncological diseases. The composition of 20-carbon fatty acids in guinea fowl breast meat is especially notable for C20:4 (arachidonic), whose levels are higher than that found in breast meat of other types of poultry, including free-range chicken in conditions similar to those found in extensive guinea fowl production. The saturated/unsaturated ratio shows mean values of 0.48 compared to the value observed in chicken of 0.45 (García Torres et al. 2009). These values also depend on both sex and variety of guinea fowl (Śmiecińska et al. 2022).

Sensory Quality of Guinea Fowl Meat

Any physicochemical meat quality study needs to complemented by a study of the sensory quality of the product. This is due to the importance of understanding the grade of acceptance of the meat by consumers.

The agri-food chain is made up of numerous links, the first being the producer or livestock farmer and the last the consumer. The growing importance given to quality aspects has been one of the most important factors of change in this chain. In any

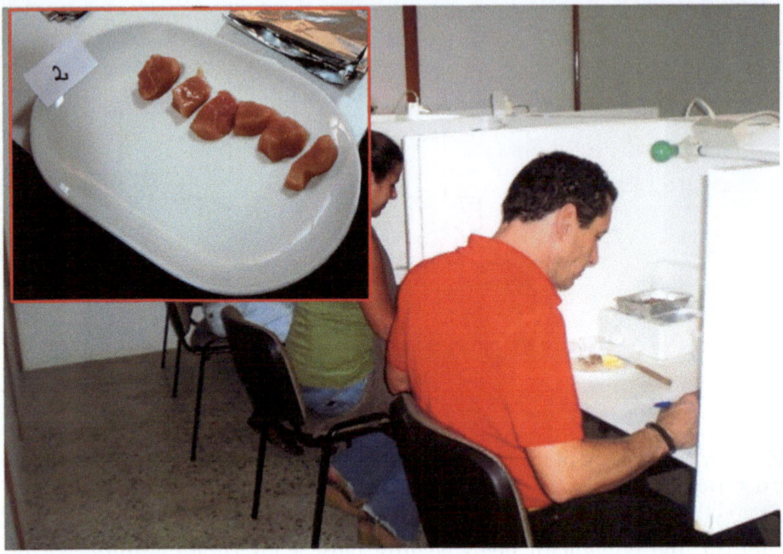

Fig. 8.10 Guinea fowl meat tasting panel

case, the value that the final consumer places on different attributes related to sensory perception will be of crucial importance for producers who focus their activity on the attributes most demanded by end users.

Sensory analysis is the scientific discipline that allows the objective and reproducible measurement of the characteristics of a product through the senses. This requires a panel of trained tasters, which means considering tasters themselves as measuring instruments (Fig. 8.10). The training is exhaustive and specific to each product being tasted, and the panel provides us with final information on the characteristics or attributes of the meat (guinea fowl in our case) as well as its degree of consumer acceptance.

The study carried out by a trained sensory panel makes it possible to evaluate, among other parameters, the intensity of poultry odour, intensity of other odours, tenderness, juiciness, oiliness, fibrousness, intensity of poultry flavour and overall appreciation.

There are differences in the sensory quality of guinea fowl meat depending on the season of the year such as tenderness, juiciness and fattiness, with breast samples showing greater tenderness in winter. Juiciness is also higher during this season, which enhances the sensation of tenderness. The same positive evaluation can be observed in the fattiness.

In summary, we can say that the quality of guinea fowl (*Numida meleagris*) meat is rich in proteins of high biological value, firm, tender and juicy, with a characteristic flavour as determined by a panel of tasters. From a nutritional point of view, it is low in fat, and its fatty acid profile determines a high percentage of heart-healthy fatty acids (Del Moral et al. 2006; Koné et al. 2022).

Microbiological Quality of Guinea Fowl Meat

The chemical composition of poultry meat has a significant influence on the growth of all kinds of bacteria, especially spoilage-producing and pathogenic bacteria. It is a good source of protein, vitamins and minerals, which, together with its water activity and pH, make it an excellent medium for microbial growth.

The bacteriological quality of poultry meat depends on different factors related to the slaughter of the animal, subsequent processing and marketing, not forgetting its rearing during its lifetime. In the different stages of poultry processing, there are increases or decreases in the microflora present in the carcasses.

One of the most important steps during the processing of poultry is scalding, as it is a critical control point and, together with evisceration, is decisive for the contamination of the carcass.

The temperature of the scalding tank is approximately 60 °C or higher to prevent the proliferation of *Salmonella* in the water. It has been shown that temperatures of 56 °C, which do not damage the epidermis, are effective in reducing its presence. Spoilage microorganisms are also affected, which may explain why birds subjected to light blanching tend to have a longer shelf life due to slower microbial growth.

After scalding, the final stage is the cooling of the carcass, after which the microbial counts on the skin usually vary from 5×10^3 to $1 \times 10^5/\text{cm}^2$ and in the abdominal cavity from $1 \times 10^4/\text{cm}^2$ to $1 \times 10^5/\text{cm}^2$. Over time, the flora increases; after 10 days, there may be between 1×10^9 and $1 \times 10^{11}/\text{cm}^2$ on the skin. This increase is accompanied by abnormal odours and a sticky slime layer.

With an initial microbial load of 10^3 germs/g, the meat lasts about 12 days without showing spoilage; if the initial load is 10^5 germs/g, spoilage starts after about 6 days.

In terms of pathogenic flora, guinea fowl meat is one of the major sources of human foodborne infection. The main causative agents are *Salmonella*, *Clostridium perfringens* and enterotoxigenic *Staphylococcus aureus*, along with some enteric diseases caused by other microorganisms, especially enteropathogenic *Escherichia coli*, *Yersinia enterocolitica* and *Campylobacter jejuni*.

The microbiological parameters required by legislation for the marketing of fresh carcasses include mesophilic aerobes, psychrotrophs, *E. coli*, *Cl. perfringens*, *S. aureus* and *Salmonella*.

In studies carried out in Spain on guinea fowl carcasses reared in the south-west of the country (Extremadura), both in intensive and extensive systems, a preservation system was established based on a treatment that reduces the initial microbial load in the guinea fowl carcasses and slows its growth for as long as possible, thereby extending the shelf life of the product and improving the handling, distribution and marketing of these carcasses. Specifically, with this treatment, reductions of between 40% and 50% in CFUs (colony forming units) were achieved compared to the initial results (López Parra et al. 2007).

Packaging of Guinea Fowl Meat for Marketing

The aim of meat production is to market it and introduce it into the catering world with all the guarantees that the product requires, both from the organoleptic and microbiological and sanitary safety. Therefore, finding a preservation system that increases its shelf life without affecting the quality of the carcasses is generally a commercial objective.

Increasing the shelf life of a product means being able to keep the product waiting to be used for a longer period, whether on the supermarket shelf or in the cold store of a restaurant.

The limiting factor in the shelf life of chilled meat is microbial growth. Chemical inactivation of microorganisms in meat includes the use of spraying or, in the case of skin-on poultry, immersion in suitable agents. Thus, for example, organic acids or their salts are used to aid the preservation of a wide variety of foods. Short-chain acids such as acetic, benzoic, citric, lactic, propionic and sorbic acids and their salts are the most used. The inhibitory effects of organic acids are closely related to the amount of undissociated acid present, as the inhibitory power of undissociated acid has been shown to be 60–100 times higher than that of dissociated acid. This is due to the undissociated acids' greater ability to cross the cell membrane; once inside the cell, they are converted to the dissociated state and exert their action. Their effects are profound, including acidification of the cytoplasm and inhibition of substrate transport, energy-producing processes and macromolecular synthesis. Generally, organic acids are most effective in foods with a pH below 5.5, although others are most active at a pH between 6 and 7.

In the case of guinea fowl meat, as it is a poultry, the treatment with organic acids is carried out by immersion. This, combined with the scalding temperatures of 50–55 °C, increases its effectiveness in reducing the microbial load of the carcass, although the degree of the reduction is variable. Its effect on pathogenic microorganisms is extrapolated to that on spoilage microorganisms. Although there is the possibility that by eliminating the altering flora it may provide an opportunity for pathogenic flora to grow, studies have demonstrated that this treatment increases the shelf life of the meat.

Organic acids differ in their effects against moulds, yeasts and bacteria. Many combinations of organic acids and other barriers are synergistic. Lactic and acetic acids and their salts have often been used as decontaminants for their antimicrobial properties in the meat industry. In general, decontamination is more effective when microbial counts are high. Under these conditions, it has been proven that the application of 3% acetic acid at 10–15 °C causes a decrease of the mesophilic aerobic load by up to 99% and increases the storage time from 18 to 21 days.

The use of lactic acid as a disinfectant allows the decontamination of poultry thighs without altering the taste, odour or texture of the product. It affects various pathogenic bacteria, inhibits spore-forming bacteria at pH 5 and inhibits the growth of acid-tolerant yeasts, and, in some cases, inhibits the formation of mycotoxins.

Acetic acid is widely used as a preservative. Its mode of action is similar to that of the other acids. Its inhibitory ability is considered better against bacteria than against moulds and yeasts.

On the other hand, it is well known that the use of cold in the preservation of perishable products such as meat, fish, milk, fruit, etc. is essential, as low temperatures inhibit enzymatic and microbial activity, which are responsible for the alteration of these products. However, many of these microorganisms remain in the food and in the environment under different resistant forms, resuming their activity when temperature conditions are favourable. Therefore, it is necessary to improve the preserving action of cold, as well as its deodorising power, since it reduces odours, but does not eliminate them.

The use of modified atmospheres to increase the shelf life of a product is not a new concept in food preservation. From 1960, with the development of new materials and technology, modified atmosphere packaging began on a large scale, and in the 1980s, this trend of protective gases reached the retail market.

A modified atmosphere is defined as a change in the gaseous composition of the atmosphere in contact with the product. Based on this definition, the following can be included:

- The atmosphere surrounding a vacuum-packed product (Fig. 8.11)
- The atmosphere obtained by injection of one or more gases
- To the controlled atmosphere of constant composition

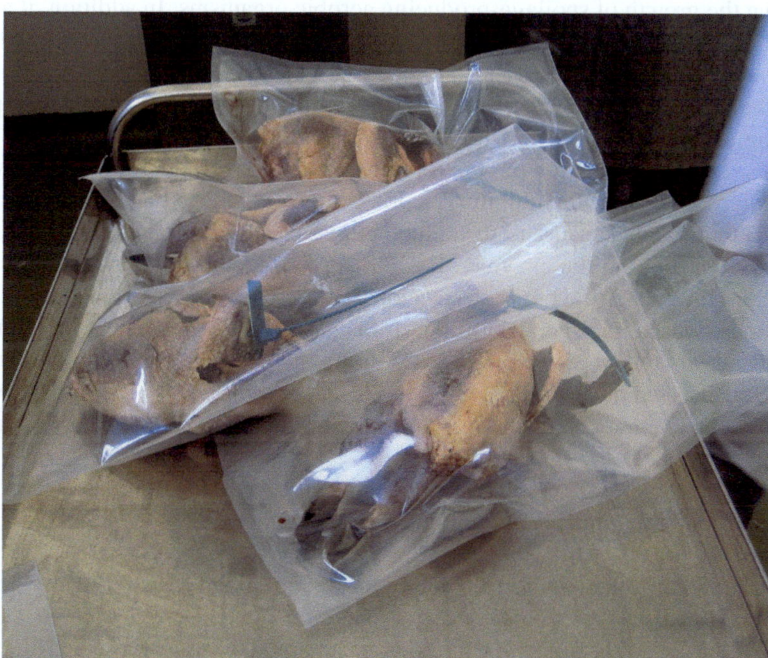

Fig. 8.11 Vacuum-packed guinea fowl carcasses

The shelf life of perishable products preserved in a normal atmosphere is limited by two main factors: the effect of atmospheric oxygen and the growth of spoilage-producing aerobic microorganisms. These factors, individually or in association with others, lead to an overall deterioration of product quality. Refrigerated storage may delay this deterioration but would not increase shelf life sufficiently to meet retail distribution requirements and point-of-sale display objectives.

Modification of the atmosphere inside the packaging, by reducing the oxygen content, significantly extends the shelf life of perishable foods kept at refrigeration temperatures.

The gases most used to achieve these atmospheres are the following:

- Oxygen. It is an odourless, colourless, tasteless gas that prolongs and maintains the colour of the meat. In modified atmosphere packaging, oxygen is eliminated or reduced to as low a level as possible, as it is one of the main spoilage agents in food. In this way, oxidation reactions that cause unpleasant tastes and odours and the growth of pathogenic and spoilage microorganisms that need it for their metabolic activity are inhibited. Exceptions are when this gas is necessary for respiration of fruits and vegetables, colour retention as in red meat or to avoid anaerobic conditions in white fish.
- Nitrogen. It is an inert gas that does not interact with food. It is not very soluble in water or fats. In modified atmosphere packaging, it is mainly used to displace oxygen, thus slowing down oxidative rancidity and lipid auto-oxidation of meat. Indirectly, it can also influence microorganisms in perishable foods by slowing down the growth of spoilage-producing aerobic organisms. In addition, it acts as a filler to prevent "package collapse" when excessive carbon dioxide dissolution takes place in food tissues.
- Carbon dioxide. Among the main gases applied in protective atmosphere packaging, CO_2 is the only one with bacteriostatic, fungistatic and insecticidal properties. It inhibits the growth of aerobic bacteria but does not delay the growth of all microorganisms. Its inhibitory action is due, on the one hand, to the decrease in pH that occurs because of the combination of CO_2 with water and, on the other hand, its action on the enzyme system of bacteria, causing damage that can be lethal. However, it favours the development of other microorganisms, such as lactic acid bacteria. The absorption of this gas is highly dependent on the moisture and fat content of the products. With foods with high moisture and fat content, such as red meat, poultry and seafood, excess carbon dioxide absorption can lead to the phenomenon known as "package collapse", especially at refrigeration temperatures.
- Other gases. Other gases such as argon, chlorine, ethylene oxide, nitrogen dioxide, etc. and mixtures of different gases are commonly used.

Other systems that enhance the action of cold include the use of ozone in cold rooms and ozonised water baths.

Ozone has a lethal effect on microorganisms and is widely used in the meat industry—in slaughterhouses, cutting and processing rooms, cold storage facilities and retail warehouses. The use of ozone for meat preservation is long established and has been applied in various ways: in cold rooms with a low concentration ozonated atmosphere for extended periods; with intermittent exposure to higher ozone concentrations; or by immersing carcasses in ozonated water baths for a set period of time. Regardless of the method, the critical factor is the time-ozone concentration combination, which must be optimised for each specific product.

Ozone is a powerful, broad-spectrum antimicrobial agent effective against bacteria, fungi, viruses, protozoa and both bacterial and fungal spores. Its inactivation mechanism is complex, targeting multiple cellular components such as enzymes, nucleic acids and cell wall or membrane constituents. Ozone kills microorganisms by breaking down the cell wall, which leads to leakage of the cell's internal contents. The susceptibility to ozone varies by microorganism: bacteria are generally more sensitive than yeasts and moulds; gram-positive bacteria are more sensitive than gram-negative ones; and spores tend to be more resistant than vegetative cells.

Studies conducted in south-western Spain (Extremadura) found that the optimal treatment of guinea fowl carcasses involved a pre-treatment with lactic acid, followed by ozonation, vacuum packaging and refrigeration. This process led to a 40%–50% reduction in colony-forming units (CFUs) compared to the initial levels. The treatment was applied to both freshly slaughtered carcasses and frozen-thawed ones, and results showed minimal differences in the physicochemical and microbiological parameters studied. Furthermore, this optimal treatment extended the shelf life of Extremaduran guinea fowl to a maximum of 10 days, significantly improving its marketability. Importantly, the treatment had no adverse effects on meat quality and even improved the texture of breast samples after 7 days of storage (López Parra et al. 2007). On the other hand, the high protein quality of guinea fowl meat, already referred to, has facilitated the design of specific peptide markers that allow product traceability to be maintained with very high reliability (Stachniuk et al. 2021). Finally, sensory analysis showed that there were no significant differences between preserved and untreated guinea fowl carcasses, which means that no odours or flavours are transferred to the carcasses after preservation treatment.

An important fact is that in sensory analysis sessions comparing guinea fowl from Extremadura, French guinea fowl, chicken and pheasant, guinea fowl from Extremadura was the most highly rated by the majority of those questioned (Del Moral et al. 2006; Rosado et al. 2011).

Competing Interests The authors have no conflicts of interest to declare that are relevant to the content of this chapter.

References

Barbut S (2024) Measuring water holding capacity in poultry meat. Poult Sci 103(5):103577

Barbut S, Leishman EM (2022) Quality and processability of modern poultry meat. Animals 12(20):2766

Bernacki Z, Kokoszynski D, Bawej M (2013) Evaluation of some meat traits in two Guinea fowl genotypes. Eur Poult Sci 77(2):116–122

Cepero R, García-Martín E, Castelló JA, Castelló F (2005) Other poultry, other poultry products: current situation. In: XLII poultry science symposium. Cáceres, Spain, pp 11–42

CIE (Commission Internationale de l'Éclairage) (2025) UNE-EN ISO/CIE 11664-5:2025 Colourimetry. Part 5: Luv colour space and uniform colour scale diagram u', v'*. UNE. Available at https://www.une.org/encuentra-tu-norma/busca-tu-norma/norma?c=N0074185

Del Moral J, Mejías A, Jiménez J, Pérez-Rojas F, Senero M (2006) Control of locust (*Dociostaurus maroccanus* Thunb.) pests with the "painted bird" (*Numida meleagris* L.), an ecological alternative or complementary procedure to the application of insecticides. Acta Científica y Tecnológica (Spain) 11:9–13

Erensoy K (2025) The effects of varying stocking densities on Guinea fowls (*Numida meleagris*) reared in a closed-barn system. I. On-farm behavioural profile and fearfulness. Appl Anim Behav Sci 283:106530

FAO & WHO (Food and Agriculture Organization of the United Nations & World Health Organization) (2024) Codex alimentarius. International Food Standards. Available at: https://www.fao.org/fao-who-codexalimentarius/codex-texts/list-standards/es/

García Torres S, López MM, Osorio C, Curbelo P, Tejerina D (2009) The fat quality of Guinea fowl (*Numida meleagris*) meat in two systems of extensive and intensive production. In: Congresos y Jornadas. Serie Producción animal-Asociación Interprofesional Para el Desarrollo Agrario, Spain, pp 625–626

Hailemariam A, Esatu W, Abegaz S, Urge M, Assefa G, Dessie T (2022) Nutritional composition and sensory characteristics of breast meat from different chickens. Appl Food Res 2(2):100233

Hoffman LC, Tlhong TM (2012) Proximate and fatty acid composition and cholesterol content of different cuts of Guinea fowl meat as affected by cooking method. J Sci Food Agric 92(13):2587–2593

Koné GA, Good M, Tiho T, Ngatta ZR, Grongnet JF, Kouba M (2022) Sensory characteristics and consumer preference for meat from Guinea fowl fed hevea seed meal or cashew nut meal supplemented diets. Poult Sci 101(12):102212

Liu Y, Yu Y, Meng Q, Wei Q, He W, Zhao Q, Tang C, Feng X, Zhang J (2022) A fluorescent pH probe for evaluating the freshness of chicken breast meat. Food Chem 384:132554

López Parra MM, García Torres S, Tejerina Barrado D, Blanco Martín T, Del Moral J (2005) Progress of results on the quality of guinea fowl meat in Extremadura. ITEA (Información Técnica Económica Agraria). Producción Animal (Spain) 26(II):816–818

López Parra MM, García Torres S, Blanco Martín MT, Osorio Membrillo C, Sanabria Tienza C (2007) Effect of preservation on the quality of guinea fowl meat. ITEA (Información Técnica Económica Agraria). Producción Animal (Spain) 28(II):672–674

López-Pedrouso M, Cantalapiedra J, Munekata PE, Barba FJ, Lorenzo JM, Franco D (2019) Carcass characteristics, meat quality and nutritional profile of pheasant, quail and Guinea fowl. In: More than beef, pork and chicken-the production, processing, and quality traits of other sources of meat for human diet. Springer International Publishing, Cham, pp 269–311

Musundire MT, Halimani TE, Chimonyo M (2017) Physical and chemical properties of meat from scavenging chickens and helmeted Guinea fowls in response to age and sex. Br Poult Sci 58(4):390–396

Rosado PE, Pérez F, Martínez N, Senero M, Del Moral J (2011) Extensive production of Guinea fowl in Extremadura. Sel Avícolas 53(3):61–66

Shoyombo AJ, Yakubu A, Adebambo AO, Popoola MA, Olafadehan OA, Wheto M, Alabi OO, Osaiyuwu HO, Ukim CI, Olayanju A, Adebambo OA (2021) Characterisation of indigenous helmeted Guinea fowls in Nigeria for meat and egg production. Worlds Poult Sci J 77(4):1037–1058

Śmiecińska K, Stępień A, Kubiak D (2022) Effect of variety and sex on the carcass and meat quality traits of Guinea fowl (*Numida meleagris* L.). Animals 12(21):2916

Stachniuk A, Sumara A, Montowska M, Fornal E (2021) Peptide markers for distinguishing Guinea fowl meat from that of other species using liquid chromatography-mass spectrometry. Food Chem 345:128810

Szalay IT, Phuong TL, Ferencz TR, Xuan KD, Kustos K, Gaal KK (2016) Assessing meat production of 3 Hungarian landrace Guinea fowl ecotypes reserved for in vivo conservation. J Appl Poult Res 25(2):139–144

Tejerina D, Lopez-Parra MM, García-Torres S (2009) Potential use of near infrared reflectance spectroscopy to predict meat physico-chemical composition of Guinea fowl (*Numida meleagris*) reared under different production systems. Food Chem 113(4):1290–1296

MIX
Papier aus verantwortungsvollen Quellen
Paper from responsible sources
FSC® C105338

If you have any concerns about our products,
you can contact us on
ProductSafety@springernature.com

In case Publisher is established outside the EU,
the EU authorized representative is:
Springer Nature Customer Service Center GmbH
Europaplatz 3, 69115 Heidelberg, Germany

Printed by Libri Plureos GmbH
in Hamburg, Germany